The Paradoxical Primate

Colin Talbot

imprint-academic.com

The moral rights of the author have been asserted.
No part of any contribution may be reproduced in any form
without permission, except for the quotation of brief passages
in criticism and discussion.

Published in the UK by Imprint Academic
PO Box 200, Exeter EX5 5YX, UK

Published in the USA by Imprint Academic
Philosophy Documentation Center
PO Box 7147, Charlottesville, VA 22906-7147, USA

ISBN 0 907845 85 1

A CIP catalogue record for this book is available from the
British Library and US Library of Congress

SOCIETAS
essays in political and cultural criticism

To Ron, Gary and Alexander
— the once and future Talbots

Contents

Preface   vi

Introduction: Paradox & Evolution   1

Part I: EXPLORATIONS

Chapter One: Beyond Rational Management   12
1. Trust Me, I'm a Guru   12
2. Paradigm Shifts and the Tyranny of Boston Boxes   14
3. The Emergence of Paradox   16
4. Organisational Paradoxes:
   Real, Metaphorical and Imagined   21

Chapter Two: A Treatise Concerning Civil Government   24
1. Human Nature and Government   24
2. Administrative Argument   27
3. Paradoxical Proverbs   29
4. Research I: Decentralising the Civil Service?   31
5. Research II: Strategy   35
6. Paradoxes, Pendulums and Tides   38

Chapter Three: The Organisation of Hypocrisy   41
1. The Evolution of Morality   41
2. White Lies and Social Hypocrisy   43
3. Paradoxes of Everyday Life   44
4. Tolerating Ambiguity: Religion in Japan   45
5. Organisational Hypocrisy   46

Part II: EVOLUTIONS

Chapter Four: The Whisperings Within   50
1. Hypothesising Paradoxical Instincts   50
2. Aggression and Peacemaking   51
3. Conformity and Autonomy   54
4. Altruism and Selfishness   58
5. Cooperation and Competition   63
6. Conclusion   67

Chapter Five: The Descent of Man   68
1. Instincts, Emotions and Intellects   68
2. Paradoxical Instincts and Individual Adaptability   72
3. Yobs, Hippies and Paradoxical Primates   74
4. Paradoxical Instincts and Social Formations:
   Fission-Fusion Societies   78

Chapter Six: Climbing Mount Paradoxical   80
1. Evolving Paradox   81
2. Modelling Paradoxical Instincts Redux   83
3. Dynamics of Paradoxical Systems   86
4. Paradoxical Instincts, Institutions and Intelligences   91
5. Towards Consilience:
   The Role of a Paradox Theory?   92

Bibliography   97

Societas: Essays in political and cultural criticism   104

Preface

I want to take this opportunity to say a 'thank you' to colleagues and one-time comrades who have helped me develop my thinking over the past three decades and more: my erstwhile comrades in the Trotskyist movement such as Tariq Ali, Robin Blackburn, Norman Geras, Ian Gough, Geoff Hodgson, and the late Ernest Mandel (some of whom have moved a long way since then and some of whom will take strong exception to what follows); David Aaronovitch — who now describes himself as a 'very rightwing Communist' (I always thought you were, Dave) — for various debates about Gramsci, Eurocommunism and even more obscure topics (mainly in Manchester Students Union Bar in the mid 1970s); Jenny Harrow, Norman Flynn, Christopher Pollitt and Michael Connolly for helping develop my academic career in a way I never imagined; and all my other colleagues and friends who have wittingly or unwittingly contributed to what follows. I have discussed some of the ideas outlined in this little book in the places as diverse as the UK, Hong Kong, Japan, South Africa, and Mexico and for the many constructive (and even some of the less than constructive) comments — many thanks. But, as usual, it is not their fault.

Perceptive readers will notice that my chapter headings are in fact the titles of a number of (highly relevant) books. I have shamelessly stolen this idea from the science fiction author Ken Macleod (another ex-Trot comrade) who used it in his marvellous book *The Star Fraction*. I hope he will not mind, and that Richard Dawkins will forgive the slight alteration to the title of one of his books (Chapter 6).

Finally, can I thank Ken Macleod for his 'fall revolution' books which still give me acute pangs of nostalgia every time I read one and combine my two life long passions — politics and science fiction.

Colin Talbot
Nottingham
November 2004

Introduction

Paradox and Evolution

This is a little book about a big subject — why humans are weird. We act in often apparently bizarre and inexplicable ways. Our behaviour seems sometimes to be utterly unpredictable at the individual level and almost as difficult to forecast at the group level. Human systems — such as stock markets, fashions and musical tastes — seem to change capriciously. This book suggests a not so novel hypothesis for how we are and a fairly novel hypothesis about how we got to be how we are.

The not so novel hypothesis is that people are paradoxical — that is they act in ways which are often contradictory, indeed self-contradictory. They can behave as war-mongers and peaceniks, avaricious thieves and altruistic Samaritans, cooperative bees and lone wolves, conformist teachers' pets and rebels without a cause. Not just different people — the *same* people can do all these things.

This idea is not so novel because it has been around a long time in various religions — everything from Christianity seeing humans as part divine and part devilish through to Chinese Taoist ideas about yin and yang in human behaviour. Maybe the religions were on to something, even if they express it in mystical terms, because there is quite a lot of thoroughly modern and scientific literature which suggest that humans do indeed behave paradoxically — from organisation theory, economics and other sources (as we shall see later).

However, treating paradoxical behaviour (and its source in paradoxical instincts) as axiomatic about humans is taking this a step further than most writers have done so far. Most social scientists have retreated into one of two camps: either adopting a 'blank slate' view of the sources of human behaviour or rather one-sided views of heritable behaviour. What is proposed systematically in these pages is that human instincts and behaviour are permanently contradictory — which is what we mean by paradoxical. Understanding this paradoxical nature is fundamental to understanding our branch of life on earth.

We share some of these characteristics with some close relatives in the primate branch of life (and even a few other animal neighbours). This is hardly surprising as we are an evolved species and didn't (contrary to what some people say) simply materialise out of thin air. The novel hypothesis is just that — humans have *evolved* paradoxical instincts. We are weird because we evolved that way. It is deeply buried in our evolutionary history and is therefore ineradicable.

This idea will certainly be attacked as another 'just so story' which seems, these days, to be an epithet readily applied to anyone else's hypotheses that you don't like. I am happy to admit this is a guess, a hypothesis or even a 'just so' story if you like. But as with all such ideas the only way to find out is to test them. First you have to actually read what evidence has been drawn upon, inferences made and conjectures conjected before you can understand where the idea comes from. Then you have to see if there's contrary evidence which successfully rejects the idea or not. Then you can call it a 'just so' story if it doesn't 'stack up'.

The reason the 'paradoxical primate' hypothesis might cause a bit of a stir in certain circles is because it leads to some pretty obvious and for some unpalatable conclusions. All sorts of utopians have built ideological schemas not about 'how the camel got its hump' but about 'how humans would behave if society were just so' (although for some reason these are not considered 'just so' stories). Some of these ideas have even been tried out. Unfortunately they didn't quite work out. The 'paradoxical primate' hypothesis suggests that such utopian schemes — if they run contrary to human contradictions — will never work out. I confidently predict this will not be seen as good news by various ideologues (of both left and right) — I do hope I am right.

So by now, if you are convinced that nurture rules, or that everyone is just a simple rational utility maximising machine, or you are just uncomfortable with illogical stuff like paradoxes you will have guessed this book is not for you. For those with more open minds (even some of the above) I hope it will convince you that there is something a bit more than a 'just so' story in these pages and it is worth exploring further. It also starts to make sense of an awful lot of stuff that we've had problems with for the past 5,000 years or so and many social scientists are still going quietly bonkers over (well, some of them anyway). But first, a little personal diversion (but it does tell you something, so bear with me).

This book — in part at least — recapitulates my own intellectual journey in trying to understand the peculiar beast we call 'human'. My personal trajectory is not at all novel: — a radical Marxist in the 1970s (my

20s), mellowing to a progressive reformer in the 1980s and (I hope) a more reflective pragmatist in the 1990s. If I were to try and sum up my trajectory it would be roughly from materialist dialectician to materialist rationalist and now to materialist paradoxicalist. (I am not sure 'paradoxi- calist' is a word — but if it isn't, it is now). Whilst I have lost some intellectual excess baggage and some comrades on this journey, fortunately not all I have acquired along the way has proved completely useless (I hope).

Even Marxism still has some useful things to say, however unfashionable that might seem in the post-communist, end of history, 'Noughties'. Dialectics as an approach is very close to paradox — but it is not the same thing and has the unfortunate tendency, through the notion of synthesis, to assume inevitable progress or sometimes regress — 'socialism or barbarism' in Rosa Luxemburg's famous phrase. What really often exists — in human systems — are cycles and fluxes, which sometimes progress (or regress) but more often just change over time in the balance between contending and contradictory forces.

The idea which proved to be a 'eureka' moment for me was the idea of paradox. Let me start by giving a very crude (and stipulative) definition — by paradox I mean 'permanent contradiction'. That is, something which appears logically impossible or inconsistent but which is nevertheless the case. This applies not just to ideas (the classical use of paradox) but to material and 'real' things, including and especially human behaviour, culture and institutions. Paradoxes are real. They are all around us in our religions, political beliefs, institutions, organisations and even inside our heads.

I am not the first to see the importance of paradox — if you manage to read the whole of this book I think you will come across paradox in many surprising places. Nor am I the first to see its importance for human organisations. Where I think this little book breaks new ground is in pulling together and synthesising a lot of work and thought already undertaken by others, but often not connected to each other. Above all, I am grateful for the work of Bob Quinn, whose ideas about paradox in organisations I discovered in about 1990 and have influenced me deeply ever since (but he is not to blame for what appears in these pages).

'Paradox' feels today like a semi-fashionable term in many fields of social science writing, but it was not always thus. We conducted a little analysis of articles in scholarly journals in the social sciences from 1970 up to date (see Figure 1). This shows that in the 1970s and 80s 'paradox' was a very sparsely used term. After 1990 there is a clear explosion in its use. In the twenty years between 1970 and 1989 the average number of articles per year was only 13.4, between 1990 and 1999 this shot up to 68

Paradox and Evolution

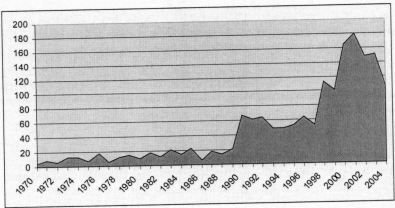

Figure 1: 'Paradox' articles in the social sciences 1970–2004

(Source: Derived from the International Bibliography of the Social Sciences.
All articles containing the word 'paradox' in Title, Keywords or Abstract.
As at 26th June 2004, so figures from 2004 represent only 6 months.)

per year and between 2000 and 2003 it rose to 161 per year. In the final year of our survey it stood at 106 for only 6 months worth of journals.

Of course this little survey may be flawed — maybe the term has simply become fashionable and therefore appears more often even though any serious exploration of the actual concept of paradox in human systems is absent from many of the articles. This is to some extent true, but even so it indicates something of a shift in perceptions and interests, but it is also true that many of these articles do have some serious considerations of 'paradox' issues in them. On the other hand, searching only on the term 'paradox' may well have missed many articles which consider the essential concept under other terminology (e.g. dilemmas, ambiguities, contradictions, etc.). This crude survey therefore probably does represent a significant shift in attention and interest towards 'paradox' as a research issue in human systems.

The watershed year seems to have been 1988. That year saw Quinn's seminal *Beyond Rational Management* (Quinn 1988) as well as the edited collection involving a range of organisation theory 'heavy-weights' (Quinn and Cameron 1988). Other management texts published that year also started to examine paradox (Harvey 1988; Pondy, Boland et al. 1988). Co-incidentally two important books dealing with paradox in group life (Smith and Berg 1988) and 'ideological dilemmas of everyday life' (Billig, Condor et al. 1988) also appeared. A popular philosophical text also appeared discussing paradox (Poundstone 1988).

Finally, it is usually a safe bet that something significant is going on when the management consultants (e.g. Price Waterhouse 1996) and management gurus (e.g. Handy 1995) catch on to a new concept. Contrary to popular belief the consultants and gurus do not usually lead this process but follow it, usually *after* initial research by academics which they shamelessly plagiarise (see Huczynski 1993 for an extensive analysis of this process). So the fact that they have now tried to stake a claim on 'paradox' suggests something is really going on. What is it?

Let's first take a small detour into the history of social sciences. The social sciences originally set out to emulate the physical sciences and discover the laws of motion of human societies. They have had poor and patchy results and some social scientists have retreated from the whole idea and sought refuge in the ultimate relativity of a socially constructed humanity.

There are, however, regularities in human social systems and in individual human behaviour; they are simply not as simple as some physical systems. Drop a leaf in a vacuum tube on the surface of planet Earth and you can predict with complete accuracy its trajectory and acceleration. Drop the same leaf in a hurricane and no physical scientist on earth can tell you where it is going to end up, or how fast and by what path it will get there. Gravity still works — its law has not been suspended — but other, very complex, forces come into play, far too complex to be precisely predicted. Each is based upon well understood physical laws but as an ensemble it becomes impossible to predict where the leaf will go. This does not invalidate any of these laws — it simply suggests that there are some limits to what is 'knowable'.

So part of the problem for social science has been complexity — even if every individual human were utterly predictable (but different), predicting the behaviour of complex social groups would be impossible. But as we now know, apparently chaotic or near-chaotic complex systems do have features we can understand and, if not predict, certainly develop some theories about how they work.

There is however a deeper problem in our approach to these issues and it is a philosophical one. Western social science is still dominated by a Newtonian world view. Despite the huge upheavals of relativity theory and quantum mechanics the view of the world we carry around in most of our heads is very mechanistic and linear. Many of us know about curious phenomena in quantum mechanics like the paradox that light can appear as waves or particles depending how you measure it but it has hardly penetrated our normal working lives. Where it has been adopted by social scientists it has been unfortunately transformed into a surreal, constructionist view of reality which alleges that light gets turned into waves or particles by the observer, so even material reality becomes socially constructed (how's that for an ego trip).

A realist understanding — difficult to comprehend but still materially based — of how the physical world works understands that the paradox of light's nature reflects deeper aspects of physical reality and is in no sense subjectively created.

Humans, we usually assume, are either one thing or another. Creative or pedestrian, aggressive or pacific, competitive or cooperative, rational or emotional, and so on endlessly. There was even a semi-popular set of psychological theories and practices known as 'personal construct theory' which suggested that bipolar contrasts are fundamental to human thinking processes (Kelly 1963).

Most social science has traditionally been constructed around the notion that if you are more of one, you must be less of the other. If you are more competitive, you must be less cooperative.

This book looks to a different, new and as yet relatively under-developed approach to understanding human behaviour at the individual, group and social levels. Paradoxical theories about human nature are not entirely new and not restricted to specific fields of social science. As we will show, elements of this approach have surfaced in evolutionary psychology, psychology, sociology, organisation and management theory and elsewhere. They have not however been systematically brought together so this is a work of fusion and, hopefully, creative synthesis that we might call 'human paradox theory'. It seeks to show why humans may have evolved a paradoxical 'human nature' and that this may indeed be a 'human universal' (although not invariable or unchanging at the individual level). These paradoxical traits or instincts in turn generate paradoxical human systems, especially human organisations — a subject which has been more fully explored — and these explorations are brought together here fully and reconceptualised in the context of the evolutionary explanation of their origins. We then go on to look at some of the implications of this approach for specific branches of the social sciences and for issues of public policy.

The book adopts a slightly odd sequence, which reflects a personal intellectual journey but has a justification as well. I happened to first develop my thinking about paradox mostly, but not exclusively, in the context of organisation and management studies. By coincidence this is also the field where it has been most developed. So I have placed the chapter on paradox in organisation and management before developing the evolutionary explanation of individual paradoxical human nature. I hope by doing so to convince the reader of the universality of human paradoxical behaviour before developing the evolutionary explanation for it, which mirrors my own intellectual journey.

It is worth saying something briefly about my own intellectual history because it helps to understand this journey and why some bits were easy and some very difficult. I was for about a decade a convinced and committed Marxist (of the Trotskyist variety). As those who have been through such intense periods of idelogical commitment will know, breaking away does not come easily. However my Marxist past both equipped me and hindered me in trying to understand paradoxical human nature. In understanding paradox it was a clear help. Many of the interesting debates within Western Marxism during the 1970s addressed a whole host of issues about dialectics which come close to (but differ in certain very important respects from) understanding notions of paradox.

The whole idea of contradiction, and moreover contradictions which co-exist despite logical incompatibility, is of course not alien to Marxist thought. Where it differs from what I now call a paradox approach is that Marxism always assumed that contradictions can be resolved by synthesis or dissolved by collapse of the system. Rosa Luxembourg's idea that the future of capitalist contradictions meant it was 'either Socialism or Barbarism' gets beyond only progressive synthesis but still does not admit that there is a 'third way' as her then social democratic contempories such as Karl Kautsky argued, long before Tony Blair latched on to the idea. Marxism acquired this legacy from Hegelian dialectics which was wholly positivist and assumed the working out of contradictions through successive syntheses towards an ultimate utopian, contradiction-free, solution. For Hegel this was an idealist working out, but Marx's great innovation was to place dialectics into the material world of developing human societies, and especially his analysis of classes and the evolving modes of production employed by successive human societies (agrarian feudalism, mercantile capitalism, industrial capitalism, industrial socialism, etc.).

The paradoxical approach does not accept this inevitable unfolding of thesis-antithesis-synthesis but instead accepts that contradictions can remain permanently unresolved. An understanding of Marxist dialectics certainly helped me understand paradox.

It is in the evolutionary part of my story that my Marxist past created the biggest problems. As a rigorously materialist approach, with none of the baggage of religion to get in the way, Marxism readily adopted Darwinian thinking about evolution. It had no trouble at all accepting humans were descended from an ape-like ancestor. Where it clashed with evolutionary theory was in the outcome — in understanding what humans are like now. Most Marxists, along with most of the rest of the utopian left, assumed that humans had evolved into something completely unique with no fixed 'human nature'. Indeed the very idea of 'human nature' became one of the chief ideological battle-grounds between Marxism and its opponents.

It was clearly impossible for Marxists to deny that actually existing humans were in many ways pretty vile creatures — robbers, murderers, rapists, destroyers. But Marxists, and other utopians, took the view of Jean Jacques Rousseau that humans are not born like that but socialised into their despicable ways by the primitive forms of social organisation they had developed (hunter-gatherer, agrarian, feudal, capitalist). Unlike Rousseau, however, most Marxists did not believe humans were born naturally good, kind and cooperative, but were simply blank slates. This assumption was essential because it explained human adaptability to nasty and brutish existence and that they could be perfected if only the correct social conditions (socialism or communism) could be created. Human perfectability and the blank slate were two-sides of the same argument — if we can create socialism we can end theft, murder, rape and destruction because humans won't want to do these things any more.

Most of us had never read Margaret Mead, but we knew by heart the arguments of her *Coming of Age in Somoa* about the vast differences in social and sexual mores in this island utopia (conveniently remote enough for us to know nothing of it other than Mead's account). Human differences, and indeed every flimsy example of cultures which appeared more utopian (mainly from hunter-gatherers, with the odd contemporary commune, free school or revolutionary episode thrown in for good luck) was called in aid of the blank slate hypothesis. This was made all the easier by the fact that many theorists of evolved human nature could be labelled (however wrongly) as fascists, eugenicists and social Darwinists and the dominant model in economics — the rational utility maximiser — was so obviously tied to right-wing ideology about selfish, un-reformable humans who just had to be 'managed'.

I must mention two interesting exceptions — the first is Norman Geras' little book on *Marx and Human Nature* (1983). Norman was a comrade in Manchester in the early 1970s and I read and re-read his book, which argued that Marx had not, despite later readings, completely rejected the idea of human nature but rather had a more sophisticated understanding of human nature plus socialisation. Whilst I agreed with Norman Geras at the time, I then had nowhere to take his ideas. They did however keep simmering away in the background of my thinking, and probably in no small part contributed to the synthesis presented here (although I expect my former comrade will not agree with much of it).

The second exception, and much more recently, is Peter Singer's *A Darwinian Left* (1999). In this Singer argues that the traditional Marxist position has been 'Darwin for nature, Marx for humanity'. In other words there is a fundamental 'break' between humans and the rest of nature. Whilst Darwinian explanations might be fine for evolution in general, and even for explaining how we came about, we as humans have now moved beyond such explanations and require a different set of explanations of how we are as we are. Singer argues that the Left have been frightened by Darwinian explanations, especially as they have been (mis)interpreted by 'Social Darwinism', eugenics and other right-wing forces. On the contrary he suggests, the Left should welcome Darwinian explanations both because they are correct and because they offer ways of fulfilling left-wing aspirations. He concludes that the Left must accept that humans are, by nature, sociable, have concern for kin, are ready to cooperate, accept reciprocal obligations, tend to live in hierarchies, recognise social status and have sex role differentiation. Culture, he accepts, has a role in 'sharpening or softening even those tendencies which are most deeply rooted in our human nature' (Singer 1999 p. 37). Most importantly Singer believes that the Left needs to understand human nature as it is, not because that implies that 'is' means 'ought', but because only by understanding how human nature is shaped by nature and culture can realistic policies be formulated which move towards desired objectives — such as a more peaceful and cooperative society.

Both these examples show that there are those on the Left who accept that Darwinian ideas have a place in understanding humans. This is not to suggest that Darwinian thinking — any more than any other scientific ideas — will not be subject to value conflicts and ideological 'interpretations' from both Left and Right. It is merely to suggest that the old divisions where Darwinian ideas were solely the property of (supposedly) right-wing thinkers and were rejected by the Left as explanations of human behaviour are fast breaking down.

This book is about combining two fundamental ideas — the idea of Darwinian evolution of human beings and the idea of paradox as a real phenomena in human behaviour and institutions — to produce a different understanding of how humans are the way we are. Whether or not it succeeds is clearly a matter for others to decide.

Part I

EXPLORATIONS

In Part I we explore the nature of 'actually existing paradoxes' in human society, culture, institutions and organisations.

We begin where I began my own journey — with explorations of paradoxes in organisations and management (Chapter 1). We go on to look at paradoxes in government and public administration (Chapter 2). Finally in this Part we explore how we deal with paradox at the social and cultural level of our society — which is mainly through that wonderful human capacity for hypocrisy, self-deception, and rationalisation, often dressed up as rationalism (Chapter 3).

Chapter One

Beyond Rational Management

As a former Marxist I was quite familiar with ideas of contradiction so it was perhaps not entirely surprising that when I started studying management and organisations in the late 1980s those theories which emphasised contradiction attracted my attention. But the most fruitful of these approaches took a very different approach to contradictions to the ones I had been used to — it suggests they are more or less permanent features of human organisations.

I was also fortunate I chose to formally study management and organisation theory rather than one of the other 'ologies'. Management and organisation theories don't seem to have a problem with seeing institutions and individual behaviours as equally important. They are, perforce, still pretty well grounded in the world of practice as well as theory which forces them to take a more rounded perspective perhaps? In contrast a lot of other social sciences had abandoned much interest in institutions and concentrated on either individual behaviour or 'culture', although fortunately institutionalism has started to come back into fashion (March and Olsen 1989; Peters 1999). Management and organisational theory has also been far more disrespectful of so-called 'disciplinary' boundaries which have hampered other social sciences that have increasingly retreated into their silos. Which is not say management and organisational theorising is not without its problems.

1. Trust Me, I'm a Guru

A few years ago on UK television one particular well-known management theorist presented what is called in the trade a 'Boston Box' (a 2x2 grid with two dimensions — these are explained further below). In this case it divided companies up by dimensions of product cost (low and high) and quality (also low and high). The other interviewee was a supermarket company manager whose main slogan was 'good food costs less at Sainsbury's'. In other words you could, paradoxically, be successful delivering high quality and low cost, which didn't fit the logic of the guru's beautifully rational model.

Now you can argue over who was right, but what was really going on here was a clash of world-views. The supermarket boss had no difficulty accepting the paradox — simultaneous existence — of high quality and low cost. It may not appear logical, but it could be done. The rational management scientist couldn't accept such arrant irrational nonsense. In his view you could simply not have two (or more) mutually contradictory goals operating at the same time. I side with the supermarket boss. The world of human systems is far more contradictory than most theorising about them allows for, mainly because humans are by nature contradictory beasts.

This contradictoriness is reflected in publishing about management and organisations, but unfortunately mostly in a rather unhelpful way. Go into any airport bookshop and you will find a bewildering assortment of management books. They are confusing for several reasons.

First, it is usually unclear what the provenance of these tomes is — are they serious pieces of research based on systematic collection and analysis of empirical evidence used to construct well-founded theories or are they merely the speculations and prognostications of self-appointed management 'guru's' or former bosses? This can usually be answered fairly easily by a quick inspection of the introduction: if there is a tell tale phrase which goes something like 'this book is based upon my years of experience as a management consultant/chief executive' then it is usually of the guru variety. This doesn't mean it's always rubbish — experience can be a great teacher and the University of Life deserves respect. Unfortunately not everyone who's 'been there and done that' has actually graduated with a first class degree. Even their personal success can be misleading. It may be purely based on circumstance rather than individual ability or the lessons they draw from their triumphs might not be the really important ones. As with the academic university it is the rigorous thinking, the learning and the ability that counts.

On the other hand, the fact that a book purports to be based on serious research or come from a bone fide academic is not always a guarantee of its accuracy or usefulness either. The rise of postmodernism, in particular, has allowed all sorts of pseudo-scientific gibberish to masquerade as 'learned' and therefore worthy of respect, even whilst rejecting the whole basis of rational science and in the words of one writer on evaluation 'abjuring objectivity and celebrating subjectivity' (Guba 1990).

The second confusing aspect of these books is their apparently contradictory messages. They can advocate strong, ruthless, leadership, or gentle enabling, empowerment and other cuddly sounding things. Recently a new book extolled the virtues of 'creative destruction' (Foster and Kaplan 2001) in explicit contrast to a best seller which advocated constructing of organisations that were 'built to last' (Collins and Porras 1994). Most confusingly of all, these differences in advice are mostly unacknowledged, or only recognised in the most crude fashion. Few and very far between are any serious analyses of the differences in evidence, theories and normative advice.

Where differences are acknowledged the most usual stratagem is what might be called the 'paradigm-shift ploy'. This consists of erecting one ideal-type model of practice and/or theory (A) and then counter-posing a different model (B) and arguing that what is occurring is a paradigm shift from A to B — usually with the reporter of this change as a breathless herald of the new age which they have just discovered.

2. Paradigm Shifts and the Tyranny of Boston Boxes

This is first a misuse of the term paradigm as popularised by Thomas Kuhn, the famous history of science theorist, in his book *The Structure of Scientific Revolutions* (1962). Kuhn coined the term to describe how scientific revolutions take place when one paradigm — a reasonably coherent set of explanation, theories, methodologies and bodies of evidence — gives way to another (e.g. from Newtonian mechanistic physics to Einstein's relativistic and quantum mechanics' probabilistic physics).

The reason why much of management writing about paradigm shifts is a misuse (or to be charitable, a different use) of the term paradigm is that they have swapped reality and theory around. In Kuhn's approach he sees a paradigm as a way of seeing the world (theory) — the underlying reality is simply there, it is not changed by the way we see it. A paradigm shift is thus a change in the way we see and understand the world, not a change in reality. The universe did not shift from being Newtonian-mechanical to Einsteinian-relativistic as soon as Einstein developed his theories or they became widely accepted. What shifted was the whole problematic or paradigm or world-view of how we saw the world operating. Arguably, in any case, both Einstein's and probably even more so Darwin's ideas have yet to fundamentally change the world view of the majority of people, as opposed to scientists, of even the developed world. Kuhn was writing about scientific revolutions, not popular ones.

In management writing a paradigm shift has come to mean a change in the way the world works (from hierarchy to networks; from public administration to 'new public management'; from personnel to human resources management; etc.). It is a change in reality itself, to practice in human organisations, not primarily to the way we see it and understand it — although in fact it is that as well, but change to reality is said to precede the new theoretical understanding.

This approach is not, unfortunately, limited to the management guru literature — it permeates a lot of academic study of organisation and management as well. The principle problem with the approach is not its misuse of the paradigm concept per se, but rather what this entails for an understanding of change in the empirical reality of organisations. If what you look for is paradigm shift then your research methods will focus on looking at dramatic shifts from one internally consistent pattern of organisational of management behaviour to another. The consequence is that all contradictory and fuzzy realities — in both the pre- and post-paradigm shift state — are ironed out. Differences between paradigms are exaggerated, contradictions within paradigms ignored, similarities between the supposedly different paradigms downplayed. It relies heavily, in other words, on either/or thinking.

A related problem is the tyranny of the 'Boston Box'. This takes the form of a 2x2 box — named after the Boston Consulting Group who allegedly popularised it — which is made by creating two intersecting dimensions on which to place an aspect of organisational or management practice. This typical 'Boston Box' matrix — which is so ubiquitous in management books that hardly anyone comments on this as a method any more — is illustrated in Figure 2. This is obviously a useful analytical tool and is so common I often joke that it isn't a serious management book/article without a Boston box somewhere in it.

Dimension Y

		High	Low
Dimension X	High	Option A	Option B
	Low	Option C	Option D

Figure 2: Classic 'Boston Box'

However it has some obvious problems. There is a rigidity and an 'either/or' over-simplicity built into this model. It forces what can be fuzzy and complex real phenomena into naive categories with spurious exactness. It is useful, perhaps, for conceptualising problems and there are certainly some simple problems to which it does apply. However, when it is used thoughtlessly (which it mostly is), it can disguise more than it reveals.

There are other ways of mapping categories but the 'Boston Box' is a useful starting point because it illustrates a feature of nearly all such approaches — their use of 'either/or' categorisations. These tend to suppress paradox and contradiction which, as we shall see, are actually fairly widespread in human systems.

3. The Emergence of Paradox

We start to explain the rise of paradox by turning to what was probably the most influential management book of the 1980s — Peters and Waterman's *In Search of Excellence* (1982). This was a publishing phenomenon but perhaps more importantly it had enormous influence. It was based on the authors' experience as senior consultants at McKinsey & Co, one of the world's leading management consultancies. It examined a number of supposedly 'excellent' companies and came to conclusions which were largely at odds with prevailing orthodoxy. They stressed a 'bias for action' as against what had become known as 'paralysis by analysis'. They asserted the importance of companies being 'close to the customer' and 'sticking to the knitting' as opposed to the fashion for mergers and acquisitions which was dominant. They stressed simple organisations, reliance on their staff, empowerment, and so on — all counter to the dominant 'Taylorian' model of management.

Critics rightly pointed to Peters and Waterman's rather dodgy methodology and were later more than happy to point out that some of their 'excellent' companies had gone badly 'pear-shaped'. But the book had tremendous resonance with real managers, including public sector managers for whom the lessons were less obvious. (In the early 1980s I was working in English local government and it was not an unusual sight to see a manager wandering around clutching a copy).

Whilst most of the commentary on and drawing of lessons from Peters and Waterman's book concentrated on the 'how to do it' bits the analysis underpinning their approach tended to get overlooked. They spent some time dismantling what they called 'the rational model' of management and organisations which is most associated with the work of Taylor and 'Fordism' (Doray 1988) but which was also associated with the planning and rational analysis fad of the 1960s and 70s, most often symbolised by the approach of Robert McNamara at the Pentagon. Rational analysis, number-crunching and planning dominated business thinking and business schools and Peters and Waterman argued strongly that this neglected the 'human' side of the enterprise. This was not especially new — the 'human relations' school of organisation theory had been around since the 1930s (McGregor 1985) — but they did give it a novel 'spin'. People, argued Peters and Waterman, are inherently contradictory:

> The central problem with the rationalist view of organizing people is that people are not very rational. To fit Taylor's old model, or today's organizational charts, man is simply designed wrong (or, of course, vice versa, according to our argument here). In fact, if we are correct, man is the ultimate study in conflict and paradox. (Peters and Waterman 1982 p. 55)

They go on to lay out what they see as the essential paradoxes in human nature (writing in an organisational and business context, of course):

> (1) All of us are self-centered, suckers for a bit of praise, and generally like to think of ourselves as winners. But the fact of the matter is that our talents are distributed normally — none of us is really as good as he or she would like to think, but rubbing our noses daily in that reality doesn't do us a bit of good.
>
> (2) Our imaginative, symbolic right brain is at least as important as our rational, deductive left. We reason by stories at least as often as with good data. 'Does it feel right?' counts for more than 'Does it add up?' or 'Can I prove it?'
>
> (3) As information processors, we are simultaneously flawed and wonderful. On the one hand, we can hold little explicitly in mind, at most a half dozen or so facts at one time. Hence there should be an enormous pressure on managements of complex organizations especially to keep things very simple indeed. On the other hand, our unconscious mind is powerful, accumulating a vast storehouse of patterns, if we let it. Experience is an excellent teacher; yet most businessmen seem to undervalue it in the special sense we will describe.
>
> (4) We are creatures of our environment, very sensitive and responsive to external rewards and punishment. We are also strongly driven from within, self-motivated.
>
> (5) We act as if express beliefs are important, yet action speaks louder than words. One cannot, it turns out, fool any of the people any of the time. They watch for patterns in our most minute actions, and are wise enough to distrust words that in any way mismatch our deeds.
>
> (6) We desperately need meaning in our lives and will sacrifice a great deal to institutions that will provide meaning for us. We simultaneously need independence, to feel as though we are in charge of our destinies, and to have the ability to stick out. (Peters and Waterman 1982 pp. 55–6)

Although this 'paradox' theme in Peters and Waterman was largely ignored at the time it did represent what turned out to be the precursor of something of a surge in interest in the whole idea of paradox in human organisations — both in the 'consultancy' literature (Handy 1995; Harvey 1996; McKenzie 1996; Price Waterhouse 1996; Cannon 1997) and in more serious academic studies (Morgan 1986; Quinn 1988; Quinn and Cameron 1988; Miller 1990). It is worth exploring a few of these.

The study which is most similar to Peters and Waterman is that in Collins and Porras's *Built to Last* (1994). Unlike *Excellence*, their research was both more rigorous and designed to compare what separated 18 successful, what they called 'visionary', companies from a matched set of 18 moderately performing companies in the same sectors. The key overall finding is that successful companies operate through paradox rather than trying to suppress it. They talk about the tyranny of the 'or' — meaning dilemmas like sticking to core values OR promoting change, having low cost OR high quality, etc. Instead, they argue that their evidence shows the successful companies manage to combine paradoxical elements such as having purposes beyond profit AND pragmatic pursuit of profits; having what they call 'Big Hairy Audacious Goals' AND incremental evolutionary progress; etc. They list 11 such paradoxically paired elements (Collins and Porras 1994 pp. 43–5) and spend the majority of the book exploring them in practice in their sample companies. Here, set out diagrammatically, are just three of their paradoxes (p. 44):

Selection of managers steeped in the core	AND	Selection of managers that induce change
Ideological control	AND	Operational autonomy
Extremely tight culture (almost cult-like)	AND	Ability to change, move and adapt

The parallels here with Peters and Waterman's 6th point cited above about individual motivation are obvious:

We desperately need meaning in our lives and will sacrifice a great deal to institutions that will provide meaning for us.	AND	We simultaneously need independence, to feel as though we are in charge of our destinies, and have the ability to stick it out

Organisations, Collins and Porras are suggesting, need to be both rigid and flexible to be successful and survive or, to put it another way, they need both continuity and change. Looking at the individuals that make up human organisations, Peters and Waterman suggest that we are motivated precisely to want BOTH belonging, stability, commitment, and individuality, change and autonomy. The links are rather obvious — the successful organisations are the ones which mesh most effectively with, and take advantage of, these paradoxical individual characteristics. As we shall see (especially in Chapter 2) these are important factors in the make up of the 'paradoxical primate'.

Probably the most sustained exploration of the issue of paradox in organisations and management has been conducted by Robert Quinn, Kim Cameron and their colleagues (Quinn 1988; Quinn and Cameron 1988; Quinn, Faerman et al. 1996; Cameron and Quinn 1999). They argue that the organisational and management literature of the twentieth century can be divided into four, mutually exclusive, models:

- the **Rational Goal** model
  (which emerged between 1900–1925);
- the **Internal Process** model (emergence 1900–1925);
- the **Human Relations** model (emergence 1926–1950);
- the **Open Systems** model (emergence 1951–1975);

The characteristics of these four models were set out by Quinn et al:

	Rational Goal	Internal Process	Human Relations	Open Systems
Criteria of effective-ness	Productivity, profit	Stability, continuity	Commitment, cohesion, morale	Adaptability, external support
Means–ends theory	Clear direction leads to productive outcomes	Routinization leads to stability	Involvement results in commitment	Continual adaptation and innovation lead to acquiring and maintaining external resources
Emphasis	Goal clarification, rational analysis, and action taking	Defining responsibility, measurement, documenta-tion	Participation, conflict resolution, and consensus building	Political adaptation, creative problem solving, innovation, change management
Climate	Rational economic: 'the bottom line'	Hierarchical	Team oriented	Innovative, flexible
Role of manager	Director and Producer	Monitor and Coordinator	Mentor and Facilitator	Innovator and Broker

Figure 3: Four Models of Management
(Source: Quinn, Faerman et al. 1996 pp. 10–11)

It is worth just spelling out a little more clearly what these models mean.

The 'rational goal' model was most closely associated with the work of Frederick W. Taylor, who invented that icon of the twentieth century — the white-coated person with a clip-board and stop-watch carrying out a 'time and motion' study. Taylor's approach concentrated on discovering the 'one best way' to do any job and then providing incentives to workers to do it. This is why it is characterised as both 'rational' and 'goal' oriented.

The 'internal process' model relates to that other great icon of the 'bureaucracy'. Max Weber and Henri Fayol have been most associated with both analysing and (especially in Fayol's case) prescribing how bureaucracies should work. The concentration is on designing hierarchies of posts, with clear rights and responsibilities, and processes and rules which govern how each task is to be completed — hence the 'internal process' model. This differs from the 'rational goal' model primarily because it focuses inwards and on process rather than outwards and on goals.

The 'human relations' model emerged in the 1920s and 30s and it focussed on people — what a later writer called 'the human side of the enterprise' (McGregor 1985). It is most famously associated with the 'Hawthorne experiments' where Elton Mayo, Fritz Roethlisberger and colleagues studied the effects of changing working conditions on productivity. They discovered what has become known as the 'Hawthorne effect' — that simply by paying attention to workers and making any changes to their conditions (almost no matter what they are) which are perceived as trying to treat them better improves productivity. Unlike both the previous models this approach has concentrated on multi-faceted human motivation — that people do not just work for money (rational goal) and that they value some 'self-actualisation' and 'autonomy' as against rigid rules (internal process/bureaucracy).

The final model is the 'open systems' approach. People such as Katz and Kahn at University of Michigan and Lawrence and Lorsch at Harvard (Lawrence and Lorsch 1969; Katz and Kahn 1978) developed a more organic, systems based approach to understanding organisations whilst Mintzberg's seminal empirical work on what managers actually do (Mintzberg 1975; 1980) established just how fluid and flexible managerial systems actually are.

Creating a 'Boston box' of this type is, as we have noted above, nothing especially innovative, although this model certainly seems comprehensive. Where the approach of Quinn and his colleagues differs radically is in suggesting that whilst these four models of how management works are radically different and contradictory they nevertheless co-exist — they are all 'true'.

As Quinn and Cameron point out in their introduction to a collection of writings by leading organisation theorists addressing the issue of paradox, the whole notion of contradictory systems had been surfacing throughout the 1970s and 80s but that the paradoxical nature of these contradictions has often been ignored (Quinn and Cameron 1988).

It is worth here discussing exactly what Quinn, Cameron and other see as 'paradox' in a management and organisational setting. It is worth quoting their definition in full:

> Paradoxes differ in nature from other similar concepts such as dilemma, irony, inconsistency, dialectic, ambivalence, or conflict. For example, a dilemma is an either-or situation where one alternative must be selected over other attractive alternatives. An irony exists when an unexpected or contradictory outcome arises from a single alternative. An inconsistency is merely an aberration or discontinuity from past patterns. A dialectic is a pattern that always begins with a thesis followed by an antithesis and resolved by a synthesis. Ambivalence is uncertainty over which two or more attractive (or unattractive) alternatives should be chosen. And a conflict is the perpetuation of one alternative at the expense of others. In precise terms, paradox differs from each of these concepts in that no choice need be made between two or more contradictions. Both of the contradictory elements in a paradox are accepted and present. Both operate simultaneously. The key characteristic in paradox is the simultaneous presence of contradictory, even mutually exclusive elements. (Quinn and Cameron 1988 p. 2)

It might be noted that this definition contradicts Morgan's equation of Taoist philosophy (which is undoubtedly paradoxical within this definition) with Marxist and Hegelian dialectics, which is clearly not (Morgan 1996 Ch 8).

4. Organisational Paradoxes:
Real, Metaphorical and Imagined

One interesting issue (to which we will return later in the book but will be briefly examined here) is the actual nature of these paradoxes. In the management literature there are at least clearly three discernible trends in interpreting the nature of paradox: the realist, the metaphorical and the constructivist approaches.

Quinn and Cameron, Peters and Waterman, Collins and Porras and many others all adopt what is generally termed a 'realist' approach. Paradoxes are real attributes of real organisations and exist independently of their having been 'discovered' by researchers. Indeed in a discussion of the problems of researching paradoxes in organisations, Quinn and Cameron suggest that much such investigation fails to recognise paradoxes because of the 'order bias' which assumes 'consistency and symmetry' whereas 'paradoxes are paradoxical' (Quinn and Cameron 1988 p. 13).

Paradoxes have however been viewed in at least two other ways, neither of which is fully compatible with this realist perspective. The first of these is the 'metaphorical' view of organisation theory. This has most famously been developed by Gareth Morgan (1996) but has also been adopted by a number of other influential organisational theory writers (e.g. Bolman and Deal 1991). In this approach a perspective like that of what we might term 'paradoxical systems' is merely an interpretive schema. It is a useful metaphor which helps us organise some aspects of extremely complex data about organisations, but it is only one metaphor amongst many and it is only a metaphor, not a description of a causal theory about how real organisations actually operate. In this perspective 'paradoxes' is merely one of a number of 'frames' which can be deployed to think about organisations, all of equal value. In Morgan's famous text on organisation theory 'paradoxes' forms one of a whole set of metaphors (mechanical, biological, cybernetic, etc.) which can all equally be applied to organisations with equally valid results (Morgan 1996).

The second construal of paradox is what might be termed a constructionist or post-modern approach, tied into a particular interpretation of several so-called 'new sciences', including paradox, chaos and complexity theories. In this view paradoxes are one amongst many and diverse discourses which 'construct' organisations. There is actually no 'real' set of theories and causal explanations waiting to be discovered which explains how 'real' organisations work because there is no objective reality. The emergence of paradoxical approaches represents, rather, a new discourse which both constructs and is constructed by individuals inter-acting in and around organisations. This particular school of thought has found its strongest expression in the works of Ralph Stacey and colleagues (e.g. Stacey, Douglas et al. 2000; Streatfield 2001; Fonseca 2002; Shaw 2002) but also features strongly in many 'popular' interpretations of the 'new sciences' in management literature.

Lest there be any confusion, my approach is fundamentally 'realist' in approach. This 'constructivist' approach is, I think, fundamentally unscientific and unhelpful, for reasons which we discuss later in this book.

The metaphorical school of thought has its uses — metaphors have great communicative power in explaining ideas and concepts which may be difficult to understand otherwise — but it has its limitations. The various metaphors Morgan explores have a dual character — they represent both various (partial) stages in the evolution of organisational and management theory and different 'frames' for thinking about organisations. In the former case they are mostly helpful for understanding simpler organisational theories which have been superseded or augmented by more modern research and theorising. In the latter case (as actual metaphors) they can be positively misleading and are pretty hopeless as tools for carrying out scientific exploration of organisations and management.

Morgan himself suggests a framing and 'reframing' approach which relies on adopting multiple perspectives or metaphorical views of organisations — a position which comes close to the constructionist/ postmodernist multiple discourses approach. Whilst it is great fun as a way of getting novice students of organisational life to think about alternative ways of viewing their subject (and the present author has used them extensively for just such purposes) their limitations as analytical tools quickly become apparent.

This chapter has sought to establish several things, but chiefly that a small but persuasive literature is emerging about the nature, extent and dynamics of paradoxes in human organisational life.

Chapter Two

A Treatise Concerning Civil Government

This chapter will examine some of the issues surrounding the question: what sort of government might best suit the paradoxical primate known as 'human'? In doing so we take a detour back to the eighteenth century and one writer in particular, Josiah Tucker.

1. Human Nature and Government

In the eighteenth century political economists — Hobbes, Rousseau, Locke, Hume, Smith and Tucker to name but a few — speculated freely on the relationship between human nature and human institutions (for a discussion of some of these authors see Berry 1986). Much of this was bound to be mere speculation for two crucial reasons — the continued dominance of religion, which the Enlightenment was slowly shaking off, and the absence, as yet, of the sufficient development of empirical and theoretical science adequate to offer alternative explanations to creationist and religious myths. Despite these drawbacks, it is interesting that so many of the eighteenth-century writers start from speculations about human nature.

Social and political scientists of the twentieth century largely succumbed to the seductions of 'the blank slate' — as a small example, a typical recent introductory political science textbook has nothing to say about these eighteenth-century debates (Marsh and Stoker 1995) whilst others touch upon them only in the most fleeting of ways (Held 1996).

The eighteenth-century debate about the nature of government and the state (in Europe and America at any rate) is dominated by the 'Social Contract' school of thought. Two of its main proponents — Hobbes and

Rousseau — took essentially the same view on the origins of the state: it grew from human necessity to impose some sort of order on society which was not human's natural state. They had, to be sure, radically different versions of this supposed 'state of nature' — for Hobbes it was war of all against all and in his famous phrase human life was 'nasty, brutish and short'. For Rousseau a more romantic humanity lived as isolated peaceful savages. For both Hobbes and Rousseau as humans associated in larger groups so they needed the state and the social contract as a way of regulating conflict.

There were, however, other (somewhat neglected) eighteenth-century voices who took a more naturally social view of humanity. Here we turn to the writings of Josiah Tucker who developed very interesting ideas about contradictory human nature and its relationship to issues of government. Tucker wrote mostly in opposition to the 'Social Contract' views of people like Locke and Rousseau, not because he rejected some of the ideas about liberty they supported but because of their excessive individualism and failure to recognise duties as well as rights. In this sense Tucker sounds like a very early communitarian of the 'Etzioni' school (Etzioni 1988; 1993; 1996). Here is Tucker explaining why government is (partly) a work of nature and not simply an artifice of humans:

> Here it is very observable, that the Author supposes Government to be entirely the work of art, that nature had no share at all in forming it; or rather in predisposing and inclining mankind to form it. The instincts of nature, it seems, had nothing to do in such a complicated business of chicane and artifice, where every man was for driving the best bargain he could; and where all in general, both the future governors and the governed, were both to catch as much as possible. For this Author plainly supposes that his first race of men had not any innate propensity to have lived otherwise, than as so many *independent, unconnected* beings, if they could have lived with tolerable safety in such a state: in short, they did not feel any instincts within themselves kindly leading them towards associating, or incorporating with each other; though (what is rather strange) Providence had ordained, that this way of life was to be so essentially necessary towards their happiness, that they must be miserable without it: nay, they were driven by necessity, and not drawn by inclination to seek any sort of civil government whatever. And what is stranger still, it seems they were sensible, that this kind of institution, called Government, to which they had no natural inclination, but rather an aversion, and whose good or bad effects they had not experienced, might easily procure advantages which they wanted, and protect them from many dangers, to which they were continually exposed, in their independent, unconnected state. All these things, I own, are strange paradoxes to me: I cannot comprehend them. (Tucker 1967 [1781] pp. 23–4).

Tucker is arguing forcefully for a view of human nature which tends towards association and creating institutions 'naturally' rather than being wholly individualist and in which 'government' is just a necessary and artificial evil. Elsewhere in his Treatise, Tucker comes close to propounding an evolutionary explanation for this state of affairs:

> We might, I say, have naturally supposed, that Government and mankind were, in a manner, coeval: and they were grown together from small beginnings, or a kind of infant State, 'till they had arrived at a maturer age; in regard to which we might have further supposed, that they became more, or less polished and improved, according as they had received different cultures from human art and industry. (Tucker 1967 [1781] p. 41)[1]

The alternative view of humans — that we are naturally individuals — dominated eighteenth- and nineteenth-century thinking through the works of Hume, Hobbes, Rousseau and Adam Smith. Frans de Waal points out that even modern philosophers start out from individualistic assumptions: 'John Rawls goes so far as to present the "initial situation" of human society as one involving rational but mutually disinterested parties' (de Waal 1996 p. 167). De Waal goes on to point that this assumes we do not descend from animals that have spent millions of years in hierarchically structured communities with strong mutual attachments — how could this be disinterested? De Waal goes on:

> How this caricature of a society arose in the minds of eminent thinkers is a mystery. Are we so painfully aware of the ancient patterns that we crave an antidote? Fairness and mutual respect do not come easily — we look at them as accomplishments, goals we fight for — and therefore, rather than admitting that we started out close but unequal, we like to give our quest for justice weight by constructing a story about how we were originally distant but equal. Like a nouveau riche claiming old money, we twist history to legitimize our vision of society. (de Waal 1996 p. 167)

Tucker's view of human nature is seemingly paradoxical — he argues that humans are naturally both selfish and socially orientated. At different places he seems to give slightly altered importance to the latter trait — sometimes he emphasises its importance whilst at others he suggests it is less strong than self-interest and it therefore has to be reinforced by cultural and institutional arrangements (not of course the phraseology he uses). But either way he sees both self-interest and socially oriented altruism as important motivating factors in human behaviour (not unlike Margolis 1982; Le Grand 2003). In this approach he differs radically from the mainstream of political economy at the time, which tended to see humans as either innately good (Rousseau) or bad (Hobbes).

---

[1]   It is, incidentally, fascinating how often Tucker — a Church of England Dean — ascribes human nature to 'nature' and 'providence' rather than the Creator. To illustrate his social approach he even, heretically, develops a scenario where there are not just Adam and Eve 'in the beginning' but a hundred pairs of Adams and Eves.

Tucker's assumptions about human nature have implications for forms of government which might best accommodate these contradictory impulses. Tucker himself spells out only some of these, mainly around the idea of mutual obligation, the duty to pay taxes (and the beneficial role of the latter if used wisely), the duty to defend one's state and the fact that many possible configurations of polity might be workable according to the culture and development of a particular state or nation. A modern author on government espouses similar views and comes to conclusions about the nature of the 'checks and balances' needed to accommodate contradictory human impulses:

> The democratic system of checks and balances assumes neither total perfectibility nor total depravity. It sees humans simultaneously as tainted by original sin and as capable of redemption. The democratic way by no means guarantees the triumph of virtue. . . . Still, the exercise of dissent and opposition tempers the delusions of power. . . . Democracy rests solidly on the mixed view of human nature, on people as they are in all their frailty and glory. (Schlesinger 1986 p. 434)

We are drawn inexorably towards Winston Churchill's famous dictum that 'democracy is the worst form of government — apart from all the others' — for a paradoxical human species, that is.

2. Administrative Argument

Public institutions reflect the paradoxes of human nature at several levels: in their overall structuring, in their management and above all in their successes and failures.

My own primary interest is in the structuring and management of public institutions. In this field there are several studies which emphasise contradictory and paradoxical aspects of the public sector, including some of my own work. It is this area I will now look at first. Later in this Chapter I will turn to wider issues about the nature of human political institutions and why democracy and the much traduced 'Third Way' may represent something broadly in tune with the nature of the Paradoxical Primate.

Before reviewing some of the work that consciously emphasises paradox, I want to briefly mention a study which uses traditional rationalist approaches in its research methods but in doing so it unwittingly highlights paradoxical systems in a striking way.

In the early 1970s a number of researchers on the field of organisational studies — known as the 'contingency' school — believed that the structuring of organisations came about as the results of a variety of external and internal factors — contingencies. A group of researchers led by Royston Greenwood set out to apply this approach to English Local

Government. Their method was fairly straightforward — they wanted to compare the strategic stance of the local authority with its internal structuring, their hypothesis being that radically different strategic stances would lead to significant internal structural differences.

It order to operationalise what they meant by a strategic stance they decided to borrow a model from the generic strategic management literature (Miles and Snow 1978). This model posits four possible mutually exclusive strategic stances (in classic Boston Box fashion):

- **Prospectors** — who seek out new strategic opportunities
- **Analysers** — who 'stick to the knitting' but keep an eye on what the Prospectors are up to and follow if it looks successful
- **Defenders** — who stick to their own area of competence
- **Reactors** — who unpredictably flip-flop between the other three styles

Greenwood et al. devised a questionnaire for local authority chief executives which included a question which allowed them to allocate their local authority to one of these four styles (the styles weren't called by their names and their descriptions were suitably neutral).

They received a good response: 200 local authority Chief Executives — over two-thirds of the total — replied. Their answers to the strategic stance question were fascinating, but not in the way Greenwood et al. expected. Half of the Chief Executives ticked the box which related to the 'Reactor' style. This was bad news for the researchers as this is a residual and inconsistent style in the original Miles and Snow approach. Moreover, it was difficult to see how it could be related to internal structures as it was, by definition, inconsistent. They adopted what must have seemed like a good approach at the time: for the purposes of analysing the strategy/structure patterns of the local authorities they decided to ignore the 50% of their sample who described themselves as 'Reactors'. In fact, from then on they ignored the issue of why 50% had described themselves as 'Reactors' altogether.

For me, especially as someone working in local government at the time I originally read their article, it was precisely the 50% Reactors finding which was the most fascinating. How could 50% of local authorities be 'Reactors' when in the Miles and Snow model these organisations were meant to be a residual category of failing, and soon to vanish, organisations? One answer was immediately obvious: more local authorities than private sector companies can be Reactors because they can. If being a 'Reactor' is a recipe for failure and swift demise in a market context it does not necessarily mean organisational death in a public sector context (Kaufman 1976).

But another question remained unanswered — were all these 50% of English local authorities really 'failing' organisations? This was in the days before the multiple performance measures of local authorities that we have today were available so there was no direct way of telling, but it seemed unlikely. I could only rely on anecdotal evidence, but it seemed to me it was inconceivable that 50% were really that bad — something else must be going on. It was not until I came across Quinn and colleagues' work a few years later (Quinn 1988; Quinn and Cameron 1988) that I found an explanation — the possibility of paradox and inconsistency meaning both possible failure and — more surprisingly — excellence. From this perspective the likelihood is that some of the 50% of 'Reactors' were indeed failing organisations but also some may have been highly successful.

3. Paradoxical Proverbs

The late Nobel Laureate Herbert Simon is most famous, perhaps, for his assault on the purist notions of neo-classical economics. Simon argued — largely from an information processing/cybernetics perspective — that the 'rational utility maximiser' model of humans so beloved of neoclassical economics was fatally flawed on the simple grounds that, paradoxically, humans always had too much information to absorb and yet not enough to really be 'rational utility maximisers'. We have too much information to be able to process because although the human brain is a marvellous information processing machine it has its limits, and the trade off between different preferences and different opportunities for satisfying them is so large that 'calculating' a 'utility function' becomes impossible. Moreover we almost never, in the real world, have sufficient information even to be able in principle — if we had sufficient computational power — to be able to calculate the actual course of events if we make particular choices. Reality is just too complex.

Instead, Simon argued, what we do is 'satisfice' using 'bounded rationality'. That is we use a combination of heuristic strategies and probabilities based on insufficient information to make a best guess at the best way of achieving 'good enough' results (Simon 1957; 1960).

Simon also had very interesting ideas about what he called 'nearly decomposable organisations' — a topic we will return to later.

Rather less well known outside of public administration academic circles is Simon's assault on what we called the 'proverbs of administration'. In a seminal post-war article Simon suggested that a lot of what classical public administration scholars such as Luther Gullick had been advocating amounted to a set of mutually contradictory proverbs (Simon 1946):

> For almost every principle one can find an equally plausible and acceptable contradictory principle. Although the two principles of the pair will lead to exactly opposite organizational recommendations, there is nothing in the theory to indicate which is the proper one to apply.

A good example of the application of such contradictory proverbs was given in their classic account of the constant re-shuffling of UK central government departments by Mackenzie and Grove:

> The debates in these cases follow patterns which tend to recur. A shake-up of the organization will give a chance to get rid of dead wood and to make a fresh start . . . But it is a pity to dislocate existing relations, familiar to all concerned, and it will take time for a new organization to find its feet. It is desirable to group similar functions together, for instance functions of research and production . . . But it is dangerous to divide producer from user. If they are separated, the user gets into the habit of stating impracticable requirements, the producer ceases to respond quickly to need. It is desirable to get rid of small Departments because they are expensive in overheads and make coordination difficult. But large Departments are slow and cumbersome . . . And so on: in isolation, each of the conflicting 'proverbs' makes good sense, but they cancel out . . .' (Mackenzie and Grove 1957 pp. 366–7).

Simon's recipe for resolving this problem and rescuing the 'proverbs' was a positive, empirical, research agenda which would establish 'what worked' in what circumstances — accepting that all the proverbs had some validity but the real problem was when were they appropriate?

Simon only examined a few such proverbs — Christopher Hood and Michael Jackson in a widely cited piece of work took this basic idea and extended it to produce a formidable list of 99 'administrative doctrines' covering 2,000 years worth of administrative arguments (Hood and Jackson 1991).

Hood and Jackson point out that Simon's research agenda remains largely unfulfilled — something Simon himself acknowledged 30 years after the publication of the 'proverbs' article (Hood and Jackson 1991 p. 20). They cite three reasons for this failure: first, the attempt to establish empirically based rules for application of contradictory proverbs had failed (the work of Royston Greenwood and the 'contingency' school discussed above was part of this general failure); second, whatever the research said, in practice policy debates on administrative reform still assumed a 'proverbial' character; and third, positive research had fallen out of fashion and been largely replaced by 'constructivist' approaches.

Hood and Jackson decided to 'take administrative argument seriously' and construct a catalogue of contradictory proverbs and their typical justifications. They argued that whether one takes a positivist (today we'd probably say 'realist') approach to administrative proverbs/doctrines or a constructivist approach it is necessary and useful to have as complete a map of the territory as possible. Whether one wants to study when and where doctrines might be most applicable or untangle the rhetoric and discourses of administrative reform such a 'dictionary' would be useful (Hood and Jackson 1991 pp. 23–9).

Where Hood and Jackson's work perhaps misses out is that, whilst it recognises that administrative doctrines are 'often contradictory', they fail to make the leap to using paradoxical explorations of how such doctrines might actually work simultaneously in practice. Despite this they do reach similar conclusions to others about the contradictory, cyclical and unstable nature of administrative systems (Hood and Jackson 1991 pp. 17–19).

4. Research I: Decentralising the Civil Service?

An empirical example of paradoxical principles in action is the 'decentralisation' of the UK civil service during the 1990s. The watershed which sparked this process was the so-called *Next Steps* report produced for the then UK Prime Minister, Margaret Thatcher, in 1988 (Jenkins, Caines et al. 1988). This report reviewed some attempts at devolving powers over finance to line managers (the Financial Management Initiative) and other reforms aimed at improving the economy (costs), efficiency and effectiveness of the UK Civil Service. The *Next Steps* offered a radical diagnosis and prescription: the Civil Service employed over half-a-million people and disposed of vast resources in delivery services but the focus of its top people was on advising Ministers and making policy, not on delivery. So, suggested the report, the Civil Service should be broken up into a number of 'agencies' — semi-autonomous bodies focussed on specific tasks with much greater decentralisation of operational management. From 1988 onwards a programme of creating agencies eventually led to around 80% of Civil Servants working in such bodies.

This was not an isolated programme — outside the Civil Service schools, hospitals, universities and many other institutions were subjected to 'decentralisation' initiatives during the 1990s (Common, Flynn et al. 1992; Pollitt, Birchall et al. 1998). Nor was it the only decentralisation within the Civil Service, as attempts were made to push personnel and financial management responsibilities further down the line in agencies and non-agency bodies alike.

If we see decentralisation (autonomy) as one pole of a paradox within public organisations, the other being control and centralisation, then what would we expect to see happening? Any strong push towards decentralisation would be met by tendencies, some latent and others expressed, back towards centralisation. Sure enough, the contra-tendencies were evident if researchers looked for them. Many unfortunately did not, caught up in the rhetoric of the reform process themselves and operating within a rationalist approach all they saw was decentralisation. Gradually, however, the rather more complex reality started to emerge.

A key piece of research was conducted by Christopher Hood and colleagues (then at the LSE). They examined the growth of regulation, inspection and audit in the UK public sector (Hood, James et al. 1999). What they found was massive growth in such bodies and institutions for regulation, inspection and audit paralleling the decentralisation process. These were and are highly centralised bodies imposing rules, penalties and other incentives on public bodies — in other words a form of re-centralisation occurring through a different route. Whilst managerial hierarchies were being decentralised, new forms of central control were being imposed externally.

However it was not just outside the structures and institutions of the public sector that centralisation was occurring. My own work in this period primarily looked at issues to do with performance regimes, first of all just in the new Agencies for a Parliamentary Committee (Talbot 1996) and later across all of government for another Committee and more generally (Treasury Committee 2002; Talbot 2002a,b). What I found about the performance regime was what might be called 'strategic centralisation and operational decentralisation'. Whilst the power of central government was being rolled back in some aspects of managing public services — mainly in the operational details — it was being greatly strengthened in other, much more strategically important areas. The power of Whitehall to dictate priorities and resource allocation was actually being enhanced by the overall changes, not diminished as most commentators seemed to think.

Even in some operational areas this was true. Studies of the changes to finance rules showed that the Treasury, the apex of the Whitehall machine, was strengthening its ability to control what spending Departments and Agencies did in total resource terms whilst allowing them some internal flexibilities within an overall package — something they had not been able to impose before the 'reforms' (Thain and Wright 1996). In personnel, another study I conducted suggested the picture was actually very complex (Talbot 1997; 2004). In one emblematic case an Agency had been forced by the Treasury to adopt a unique pay and grading structure which it did not want just so it could be seen to be 'devolved' in its arrangements. A very strange sort of decentralisation indeed! In fact, the Treasury still operated many hidden controls over Agencies, whilst publicly claiming a 'hands off' approach.

Christopher Pollitt and I, together with many international colleagues, have extended the work on creation or reform of 'agency' type bodies and decentralisation to many countries and found in every case that the trends and counter-trends in decentralisation and 'autonomisation' have been far from simple (Pollitt and Talbot 2004; Pollitt, Talbot et al. 2004).

New Zealand, for example, was widely seen during the 1990s as the emblematic example of what became known as the 'New Public Management' (NPM) reforms, of which decentralisation formed but one part (Aucoin 1990; Pollitt 1990; Hood 1991). It had the most explicit, the most radical and the most comprehensive reform programme (Boston, Martin et al. 1996).

The reality of the New Zealand reforms are starting to emerge and the picture they portray is by now familiar: very radical decentralisation reforms were only partially and unevenly implemented and new controls emerged through various formal and informal mechanisms. Richard Norman conducted extensive interviews with those at the 'frontline' of the reform process and concluded that what was really happening was rather more complicated than the 'official story' would suggest. For example, on the issue of managerial freedoms from central controls (a key plank of the reforms):

> One CE thought the Treasury had never really loosened the apron strings, continuing to take a line-by-line interest in budgets. After discussing the accountability requirements surrounding the agency, the CE commented 'you would wonder if there was any freedom left'. This feeling of being surrounded by accountability pressures was echoed by a policy manager, who commented that freedom was tinged with a 'fair amount of looking over your shoulder'. One CE thought freedom had to be constantly guarded against the tendency of central agencies to seek more controls than were provided for in legislation. Weaker chief executives were more likely to be targeted than strong chief executives and lose their freedom to act. (Norman 2003 pp. 84–5)

Norman concludes that:

> While important elements of the 'freedom to manage' model remain, freedom has less of the heady independence that was experienced in the late 1980s and early 1990s. Freedom to hire and fire staff within broad delegations, given to chief executives, remains a distinctive feature of the model. So does the ability to vary the range of inputs, although this is constrained by heightened awareness of the potential for criticism by the media or MPs a web of case-based experience creates the new boundaries. Core

freedoms remain, but surrounded by many hard lessons about spending and actions likely to generate controversy. (pp. 94–5)

In trying to make sense of this and other data from his research, Norman draws heavily on the paradox perspective suggesting that only this approach explains the many contradictions evident in the actuality of the New Zealand reforms.

Arguments over public administration reform which highlight one aspect of changes — e.g. centralisation — rarely capture the complex realities of what is happening. Hebert Simon made a similar point about the centralisation/decentralisation issue half a century ago:

> Consider the term 'centralization.' How is it determined whether the operations of a particular organization are 'centralized' or 'decentralized'? Does the fact that field offices exist prove anything about decentralization? Might not the same decentralization take place in the bureaus of a centrally located office? A realistic analysis of centralization must include a study of the allocation of decisions in the organization and the methods of influence that are employed by the higher levels to affect the decisions at the lower levels. Such an analysis would reveal a much more complex picture of the decision-making process than any enumeration of the geographical locations of organizational units at the different levels.
>   Administrative description suffers currently from superficiality, over-simplification, lack of realism.' (Simon 1946)

If we take out the words 'field offices' and replace them with 'agencies' then we get an almost perfect description of some of the problems with much of the analysis of the UK's 'agencies' reforms described above. In the Home Office, for example, the Immigration and Nationality Department (IND) remained — in Simon's words — a 'bureau of a centrally located office'. But IND received almost all the same decentralised powers as the Home Office's 'agencies (Prisons, Forensic Science, Fire Service College and Passports Agency) without becoming an 'agency' as such (Talbot 2004).

Moreover, even when there is genuine change (as opposed to just rhetoric) the likelihood is that opposite tendencies will manifest themselves in some form or other — even if not immediately. The rhetoric of reform is almost always one-sided. 'Let's decentralise' is a lot easier to project as a slogan than 'let's change the balance of the competing pressures for centralisation and decentralisation a bit more and in specific ways in favour of the latter'. Moreover the predominant rationalist ways of thinking about such issues do not easily permit or facilitate more nuanced paradox-based explanations. As we will see later, this often leads to what one academic calls 'organisational hypocrisy' — i.e. the rationalisation of a contradictory reality into a non- contradictory 'story' (Brunsson 1989).

Researchers who try to force paradoxical systems into mutually exclusive Boston Boxes almost always miss the contrary trends and tensions. They tend either to say 'everything's changed' or 'nothing's changed' because they only see one set of evidence. Their very methods for investigating change tend to preclude seeing the contradictions, because they only ask for 'either/or' answers. For some this is just an unfortunate mistake, for others it borders on negligence. Those who extol — or indeed oppose — the latest supposed 'paradigm shift' are usually the worst offenders. Sometimes this is done purely for personal gain — being the first to spot and extol the virtues of the latest 'paradigm shift' can gain many academic 'brownie points'. Sometimes there are wider political or ideological motives. Both the opponents of Whitehall decentralisation and its critics agreed on one thing — it *was* happening. They merely disagreed over whether it was good or bad thing. Neither side had any interest in saying 'well, it's not quite like that — it's a lot more complicated and there is some decentralisation and some centralisation going on'.

5. Research II: Strategy

Another empirical example of paradox in human systems can be seen in the way in which big decisions get made in public organisations. The issue of how people make big strategic or policy decisions in the public sector has been a key topic for over half a century. It has not always been called 'strategy' or even 'policy' and the two terms are sometimes used interchangeably (e.g. prior to the 1960s in business big decisions were usually called 'business policy').

During the early 1990s — as a former public sector manager turned academic — I was especially interested in the issue of strategy in the public sector. If there is a distinction between strategy and policy it is probably that the latter is 'governmental' whereas 'strategy' is more tied (usually) to a specific organisation or institution. Moreover 'strategy', used in this way in relation to organisations, has some useful parallels in the private sector. The sort of issues that concerned me were just what is the scope for making strategy in public organisations? More specifically can public sector managers and (non-elected) leaders actually make strategy as opposed to just implement policy? And if/when they do, are their different approaches at play as Royston Greenwood's 'contingency' work on strategy and structure, discussed above, assumed?

My first step was to look at the generic literature on the nature of strategic decisions and management. As with much of the literature on management there was a bewildering array of theories and research and an awful lot of flimsily based advice. Strangely, most of the contradictions and differences of opinion in this literature was simply ignored by writers who seemed myopically concerned with their own pet theories or research. There were at the time a few honourable exceptions — most notably the work of Henry Mintzberg but also others (Whittington 1993; Mintzberg 1994) — of people trying to construct categorisations of the various theories and/or practices of strategy in organisations. None of these seemed particularly helpful by themselves for my purposes so I decided to construct my own, borrowing heavily on the 'paradox' approach of Bob Quinn and his colleagues.

It is worth mentioning at this point that as a scholar of both general management and public policy I quickly found that there were parallel but rarely linked debates going on in the two fields with often very similar underlying issues being contested. It was not the first or last time that I have come across examples of even closely related social science fields being almost totally ignorant of each others work.

The model I constructed attempted to synthesise the main strands of strategic management theory and practice into a classical Boston Box. The two dimensions I found in the literature revolved around two debates which seemed to be fundamental and underpinned many of the more detailed issues which were being discussed and researched.

The first of these was whether or not strategic decisions are rational or something else. There was a very strong 'rational' school of thought, exemplified by the writings of the doyen of strategic planning Igor Ansoff who had produced the most amazingly elaborate models for conducting rational planning processes (Ansoff 1968). In the policy field, there were likewise those who staunchly defended the rational approach — for example Steve Leach's sterling 'defence of the rational model' (Leach 1982). (See also for example Carley 1980; Rosenhead 1989). Herbert Simon's modified form of 'bounded rationality' and 'satisficing' should also probably be put into this camp (although his ideas embrace some of the non-rational approaches as well).

Ranged against these were a variety of critics who pointed to fundamentally non-rational processes in big decision making. These ranged from proponents of the 'great man' theories of leadership and their ability to make intuitive judgements (for a classic personal account see Iacocca and Novak 1987), through to those who saw big shifts coming not from single decisions but from multiple sources and emerging incrementally. In strategy Mintzberg and James Brian Quinn are the best known proponents of such approaches (Quinn 1980; Mintzberg, Ahlstrand et al. 1998) whilst in the policy field the 'incremental' school is best known through the work of Lindblom (Lindblom 1959; 1980) and in the UK of Howard Elcock (Elcock 1991). (For an epic debate on the rational versus 'emergent' approach see Mintzberg 1990; Ansoff 1991; Mintzberg 1991).

The 'incrementalists' tended to lump together the processes by which decisions got made (complex decisions structures, diffuse or segregated power, political bargaining and compromises, multiple stakeholders, etc.) and the resultant decisions (usually incremental).

The second major dispute was between those who saw strategy as big changes (whether brought about by rational planning, inspirational leaders or other means) versus those who saw strategy as a more incremental or 'emergent' process (Quinn and Mintzberg respectively) in which big shifts were often only visible in retrospect as an accumulation of smaller decisions. The work of Miles and Snow (mentioned above) explores these issues, suggesting different organisations adopt different approaches. Their Prospectors, Analysers, Defenders and Reactors represent positions on a spectrum from large, radical, change through to no change at all or even contradictory changes (Reactors) (Miles and Snow 1978). Many of the other participants in the rational/non-rational dispute are the same as in big change/incremental change debate, but the issues are not identical.

Putting the arguments together gives us a classic Boston Box matrix which highlights the four main trends in strategic management theory and practice which were called 'strategic modes' (see Figure 4):

		Strategic Change	
		Incremental	Radical
Decision Process	Rational	**The Learning Mode** Logical Incrementalism and Learning Organisations	**The Planning Mode** Strategic Planning and Management
	Non-Rational	**The Political Mode** Incrementalism, Organisational Political and Cultural Systems	**The Visionary Mode** Visionary Leadership, Entrepreneurialism

Figure 4: Strategic Content & Process Model

(Source: Talbot 1995)

There is nothing especially dramatic about this synthesis although it does capture much of the debate in a single framework. What is really novel is the way it was used. Instead of assuming that these obviously contradictory strategic 'modes' would be mutually exclusive (which

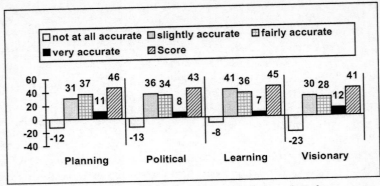

Figure 5: Strategic modes — tendency towards each

N=644, percentages of respondents — 'not at all accurate' responses
have been given negative score for illustrative purposes.

(Source: Talbot 1995 p. 272)

they logically ought to be) I assumed that they would be paradoxical,
that examples of real organisational practice would show that two or
more such modes would be operating in the same organisation. So, as
part of a much larger survey questionnaire about strategic practices in
organisations I asked UK public managers a set of questions which per-
mitted them to give logically inconsistent answers. This is in contrast to,
for example, the Royston Greenwood survey in which Local Authority
Chief Executives had to choose between one of four alternatives.

The survey results (from 644 senior managers across the UK public
sector representing almost as many organisations) were fairly conclusive (see Figure 5). The 'scores' for each of the four strategic modes were
Planning 46; Politics 43; Learning 45; and Visionary 41 — in other words
most organisations had two, three or even four strategic modes operating simultaneously. Although Greenwood's survey asked different
questions to a smaller set of public managers, their finding of 50% 'Reactors' (inconsistent strategic stance) seems remarkably similar to these
results. The evidence seems pretty conclusive that strategy in public
organisations in the UK, at least according to their senior managers, is
largely a paradoxically inconsistent process rather than a rationally
consistent affair.

6. Paradoxes, Pendulums and Tides

A number of writers on public administration, at different times and
places, have addressed the issue of how contradictory pressures in pub-

lic administration play out in practice. The paradoxical pressures they identify are not always identical and some language used varies, but the underlying concepts of mutually contradictory but always present pressures are strikingly similar. So is the language used to try and depict the results of such pressures and one of the best such descriptions comes from an Australian academic, Richard Spann:

> As in fashion, skirts go up and down, ties narrow and widen, so in Public Administration there are alternations — the economies of large-scale are preached at one time, to be succeeded by the gospel that 'small is beautiful'; periods of administrative pluralism are followed by ones of integration when semi-independent bodies become suspect . . . ; there is oscillation between the demand for functional rationality and for a holistic approach to clienteles; between the desire to politicise and to depoliticise public administration. *Sometimes tendency and counter-tendency are present simultaneously.* (Spann 1981, quoted in Hood and Jackson 1991; emphasis added).

Spann suggests that sometimes the flows between paradoxical poles exist sequentially — e.g. from specialisation to holism and back again. A very similar point is made by Guy Peters:

> The simplest explanation [of contradictory reform efforts] is that reform is cyclical and when government organizations go too far in one direction (e.g. centralization), there is a need to swing back in the other (e.g. decentralization). There are a number of such dualities that comprise a good deal of reform activity. (Peters 1998)

A good example of such cycles was work by Christopher Pollitt on UK central government departments in which he showed how there had been a constant cycling between moves to create 'mega' Ministries incorporating many functions through disaggregating these into smaller more focussed Ministries and back again (Pollitt 1984).

An even bigger example is the work of Paul Light on the 'tides of reform' in US federal government from 1945–1995 (Light 1997). He identifies four re-current reform types:

- **Scientific Management** — focus on efficiency, structure and roles
- **War on Waste** — focus on economy (cost savings), reviews, audit, inspection
- **Watchful Eye** — focus on fairness, rights, scandals
- **Liberation Management** — focus on performance, evaluations, outcomes

Light conducts the most meticulous examination of federal reform programmes over this 50 year period and calculates the efforts and results of each philosophy of reform and its relative standing compared to the others. In a summing up he suggests four phases of reform over

40   *A Treatise Concerning Civil Government*

time — the 'Glory Days' (1945–1960); the 'Great Society' (1961–1968); the 'Reassessment' (1969–1980); and the 'New Proceduralism' (1981–1994). Over these periods there is clear changing of emphasis in reform efforts, with 'Scientific Management' gradually receding and being displaced by 'War on Waste' and 'Watchful Eye' and the newcomer from the late 1960s — 'Liberation Management'.

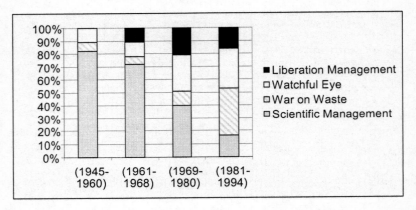

(Source: Derived from figures in Light 1997 p. 116)

As Spann suggests above 'sometimes tendency and counter-tendency are present simultaneously' — an example I gave above was of simultaneous strategic centralisation and operational decentralisation in the UK Civil Service. Light's work suggests that by the 1990s all four of his reform tendencies — despite their contradictory emphases — were simultaneously significantly present in US federal reform programmes.

There are other writers who have likewise suggested the contradictory nature of public administration and of reform processes within it — Guy Peters has addressed what he calls the 'antiphons' of public administration (Peters in Peters and Savoie 1998) and in a perceptive article Peter Aucoin writes about the 'paradigms, principles, paradoxes and pendulums' of public management reform (Aucoin 1990).

Chapter Three

The Organisation of Hypocrisy

Why are humans so consistently moralistic and hypocritical? Everywhere humans have had moral codes by which everyone is supposed to behave. These have changed over time, but every human group seems to have them even when they are only oral traditions. But everywhere (and when) we seem to simultaneously espouse and break these self-created rules.

1. The Evolution of Morality

Some philosophers, such as John Rawls in his famous *A Theory of Justice* (1971), still start their analysis of morality from the lone individual (as did Rousseau, Hobbes and many others) rather than from real humans who evolved in real social groups and not at all as isolated individuals. In this chapter we will try to analyse why humans create, and espouse, moral systems and why we also break them so consistently. (The reasons, I would suggest, are essentially the same in both instances — our fundamentally paradoxical natures.) Let us start with a brief examination of how and why we develop moral codes in the first place, and what form these take.

There is, of course, a vast literature on the development of morals going back 3,000 years or more into human history. We are not reviewing that here, just the part of literature which links morals to human evolution. Darwin himself believed that:

> At all times throughout the world tribes have supplanted other tribes; and as morality is one element in their success, the standard of morality and the number of well-endowed men will thus everywhere tend to rise and increase. (cited in de Waal 1996 p. 23)

James Q. Wilson, in his book *The Moral Sense*, argues that humans are social animals, 'who struggles to reconcile the partially warring parts of

his universally occurring nature — the desire for survival and suste-
nance with the desire for companionship and approval' (Wilson 1997
p. 123). Wilson argues that to say 'there exists a moral sense (or, more
accurately, several moral senses) is to say that there are aspects of our
moral life that are universal' (ibid p. 225). However, so far those opposed
to such views have largely been in the ascendancy (he puts it thus: 'Rela-
tivists 10, Universalists 1').

The reason that such 'moral universals' have been so elusive is
because, he suggests, people have been looking for the wrong things.
Moral dispositions rather than specific moral rules are what are universal, and the same dispositions can be represented in a wide variety of
different specific rules. Wilson believes we are indeed naturally self-
regarding but also other-regarding:

> Our moral sentiments are forged in the crucible of this [child–mother]
> loving relationship and expanded by the enlarged relationships of fami-
> lies and peers. Out of universal attachment between child and parent the
> former begins to develop a sense of empathy and fairness, to learn
> self-control, and to acquire a conscience that makes him behave dutifully
> at least with respect to some matters. Those dispositions are extended to
> other people (and some other species) to the extent that these others are
> thought to share in the traits we find in our families. (Wilson 1997 p. 226)

But humans are not always empathetic — we kill each other in
depressingly large numbers. Wilson suggests that 'one rather paradoxical answer' to this contradiction is 'that man's attacks against fellow man
reveal his moral sense because they express his social nature' because 'it
is the desire to earn or retain the respect and goodwill of his fellows',
however distorted that may be (ibid p. 227).

In *The Moral Animal* Robert Wright (1995, Chapter 18) summarises the
arguments as to how these initial moral dispositions expand to encom-
pass larger and larger sets of humans. At the inspirational end of the
spectrum, Wright cites the work of philosopher (and animal rights
champion) Peter Singer who argues in his own book *The Expanding Circle*
(Singer 1983) that through the evolution of reciprocal altruism and
human communications through language, we developed the habit of
defending our actions as 'disinterested' and 'moral' out of self-interest
but this gradually 'took on a life of its own' (cited in Wright 1995 p. 372).
Wright suggests a more cynical explanation might simply be that the
idea of expanding the circle of those included in 'reciprocal altruism'
from kin, to clan, to nation, etc might just be that those, usually religious,
'sages' doing the urging are thereby serving their own self-interests by
expanding their potential flocks (ibid p. 373).

Wright also stresses that moral injunctions have evolved, related he
suggests to the circumstances. Thus in hunter-gather villages and later,

ideas about the justifiability of revenge ('an eye for an eye' retributive type morals) predominated but as human social groupings grew larger, denser and more settled, and governments began to be established, these receded to be replaced by more jurisdictional morals (ibid p. 374).

Frans de Waal likewise suggests that moral codes might evolve as circumstances change. Writing about the affects of over-crowding on human behaviour, he argues that:

> Inasmuch as the balance between individual and community values affects moral decisions, morality is an integral part of the human response to the environment, and an important counterweight to the social decay predicted by crowded conditions. Adjusting the definition of right and wrong is one of the most powerful tools at the disposal of Homo Sapiens, a species of born adaptation artists. Morality is not the same during war and peace, or during times of plenty and scarcity, or under high or low population density. If certain conditions persist for a long time, the entire moral outlook of a culture will be affected. (de Waal 1996 p. 201)

E.O. Wilson argues that humans have acquired 'unilaterally altruistic genes' by group selection and these will be countered by their opposites acquired by individual selection. 'The conflict of impulses under their various controls is likely to be widespread in the population.' This will produce what Wilson calls 'moral ambivalency' which might be reinforced by ambiguous moral codes (Wilson 1975 p. 288). How do humans cope with ambiguous, shifting, moral impulses? We deal with it through various forms of hypocrisy, to which we now turn.

2. White Lies and Social Hypocrisy

Whilst I was driving up through northern France towards England about a decade ago I heard a remarkable BBC radio interview which has stuck with me ever since. The Chilean novelist Isabel Allende was being interviewed about her life and work. She told a wonderful story about herself. As a child she had been an inveterate liar. When she went to school she would make up all sorts of lies about what happened at home. Eventually she would be discovered in some outrageous deception and punished.

At home she would tell lies about teachers and classmates. When her irate parents visited the school the truth would come out and again she would be punished. In the interview she went on to say, 'so when I grew up I wrote all these lies down and they gave me the Nobel Prize for literature.'

Humans all aspire to be honest (or at least they say they do) but they also delight in being lied to — told stories which are patently untrue. Our bookshops and cinemas are full of such lies.

Another small illustration — former US President Jimmy Carter's mother was, so the story goes, being interviewed by a particularly pushy female reporter out for a good story. The reporter asked if it was true that Jimmy's mother brought him up never to tell a lie and that she also had never lied. Mrs Carter said that was correct. The reporter kept pressing her on whether she had ever told a lie and eventually, in exasperation, she admitted she did sometimes tell 'white lies'. The reporter pounced and asks for an example. 'Well', said

Mrs Carter, 'when you arrived this afternoon I said I was pleased to see you.'

We comfort ourselves with the notion of 'white lies', lies told for a noble or social purpose, but do we really have an objective way of telling the difference? Our attitude to lies and honesty — in just about all cultures — is ambiguous and paradoxical. Almost by way of explanation of the above story (although it wasn't written in this context) the famous Polish philosopher Leszek Kolakowski, wrote that:

> there is often a thin line between falsehood and such social virtues as discretion and politeness, but we would all admit that without them life in society would be much worse than it is: far from breathing the pure air of truth, we would stifle in a world of boorishness. (Kolakowski 1999 p. 27)

Kolakowski goes on to argue that we usually make a distinction between lying to protect oneself or others and lying to obtain unfair advantage, whilst recognising that there is absolutely no hard and fast way of defining the difference between the two. The study of deceit amongst animals and humans is now quite developed, with a number of interesting studies having been published (Brunsson 1989; Bailey 1991; Giannetti 1997; Grant 1997). Ruth Grant ascribes hypocrisy in politics as due to the particular nature of power and dependency relationships between conflicting interests:

> … hypocrisy and politics are inextricably connected on account of the peculiar character of political relationships. Political relations, ordinarily understood as power relations, can just as readily be conceived as relations of dependence. They are dependencies among people who require one another's voluntary cooperation but whose interests are in conflict. In such a situation, trust is required but highly problematic, and the pressures towards hypocrisy are immense. Because political relations are dependencies of this sort, hypocrisy is a regular feature of political life, and the general ethical problem of hypocrisy and integrity is quintessentially a political problem. (Grant 1997)

3. Paradoxes of Everyday Life

In 1988 (the same year as Quinn's *Beyond Rational Management*) a group of social psychologists published an interesting work looking at what they called the 'dilemmas of everyday life' (Billig, Condor et al. 1988).

Firstly, they point out that a great deal of rhetoric draws upon a well of mutually contradictory dictums — very similar to the 'administrative arguments' discussed by Hood and Jackson in Chapter 3. They go back as far as Francis Bacon in the sixteenth century who gave many examples of such contradictory assertions, such as 'silence is fermentation of thought' versus 'silence, like night, is convenient for treacheries' (cited in Jardine 1974 p. 224).

Secondly, they give some examples of the modern equivalents of Bacon's antithetical aphorisms drawn from everyday 'common sense' expressions, such as:

Absence makes the heart grow fonder	Out of sight, out of mind
Nothing ventured, nothing gained	Look before you leap
Many hands make light work	Too many cooks spoil the broth
Charity begins at home	Love thy neighbour

(Derived from Billig, Condor et al. 1988 pp. 16–17)

So far so good, but it at this point we start to part company. Our social psychologists see these dilemmas as purely 'ideological' in character and indeed part of the very mechanism of thought. They are primarily conflicts over values and beliefs, rather than over real social relations — in other words they adopt a social constructionist approach rather than a realist one.

4. Tolerating Ambiguity: Religion in Japan

In his work comparing national cultures, Geert Hofstede[1] has as one of his dimensions 'uncertainty avoidance', which is avoidance of ambiguity. Japan scores highly on this dimension, 97 points in Hofstede's indices, and rates only 7th in the world for 'uncertainty avoidance' (Hofstede 2003 p. 500). So here is a remarkable paradox — Japan has one of the most

---

[1] Hofstede, I should point out, is firmly in the 'relativist' camp when it come to human behaviour. In the nearly 600-page 2nd edition of his famous book comparing national cultures human nature does not get a look in. Nevertheless his empirical work is pretty unique and very interesting.

ambiguity tolerant attitudes to religion in the world (something Hofstede does not explain).

In the early 1970s the Japanese Ministry of Culture decided to undertake a survey to find out the religious affiliations of its people (Hori, Ikado et al. 1972). A survey was duly conducted. The results were rather surprising. Total religious affiliation added up to about *one and half times the Japanese population*. How could this be? Well at one level the explanation is simple. The Japanese have no difficulty getting married in Catholic Church and then going to a Shinto Shrine for a blessing ceremony, or the other way around. This is not because a Catholic is marrying a Shintoist — both the couple likely embrace aspects of both faiths. Religion is accepted as a deeply personal and idiosyncratic choice amongst the Japanese and no-one finds it in the least remarkable if their compatriots choose a seemingly illogical, and to Western eyes bizarre, combination of belief systems. Tolerance for ambiguity is extremely high in this area of Japanese life, if not in other areas — as Hofstede's work, which focussed mainly on organisational settings, suggests. But other work suggests that it is within organisations that we operate at our most hypocritical — feigning rationality and coherence whilst tolerating in practice ambiguity and contradiction.

5. Organisational Hypocrisy

Nils Brunsson has carried out the most sustained critical analysis of what he calls 'the organization of hypocrisy' (Brunsson 1985; 1989; Brunsson and Olsen 1993).

Brunsson argues that organisations are subjected to contradictory environmental demands — both about what they should produce and how they should produce it. Different professional groups inside and outside the organisation (e.g. lawyers, accountants, etc.) have competing views about how it should be run. These views are not just different; they are 'contradictory and inconsistent' and:

> difficult or impossible to combine . . . Some norms may call for a centralised organization and others for decentralization. Some groups may demand democratic management processes, others authoritarian. It may be impossible to combine the customers' product requirements with the related environmental rules. And there may not be enough money to pay all the groups which make economic demands on the organization. (Brunsson 1989 p. 8)

As organisations face increasingly inconsistent demands from their environments they are also subjected to pressures to give rational accounts of how they work; accounts that assume that such conflicts can either be reconciled or avoided. He comments that the 'vast literature concerned with "good" organizational management is chiefly concerned with getting the job done', whereas his approach is to examine how organisations cope with inconsistent norms and demands.

Brunsson concedes that 'theories emphasising coherence and consistency' in organisational life do tell us something about what goes on in organisations but they fail to explain vital parts of these issues. Basing himself on qualitative, and detailed empirical studies within organisations, Brunsson concludes that organisations operate on three levels: talk, decisions and actions. At each level the organisation seeks to satisfy some demands on it but these vary across levels. These conflicting sets of 'talk', 'decisions' and 'actions' are then rationalised into a supposedly consistent 'story' about what is going on — usually by the simple expedient of ignoring the inconsistent bits (Brunsson 1989 see Chapter 2).

These ideas apply to public policy as well as organisations and the example of them in practice which I usually give is the Conservative governments' reforms of the health service in the early 1990s — the introduction of the so-called 'internal market'. At the level of 'talk' the White Paper which introduced the reforms — 'Working for Patients' — emphasised that the reforms were supposed to offer much greater choice for patients.

At the level of 'decisions' however the reforms actually provided for choice by the 'purchasers' of health services, who would be the District Health Authorities (DHAs) and the General Practitioner 'Fundholders' (GPFHs). They would 'buy' services from the newly created NHS Trusts. Patients were to be 'represented' by these actors, rather than having any direct voice or choice.

Finally, at the level of 'actions' many of the details of the reform process actually militated against even DHAs and GPFHs making free choices — various practical, financial and logistical constraints meant that in most cases they had to 'buy' services from the same providers as they always did. In practice, this meant that where GPs could have referred patients to a number of hospitals before the reforms, afterwards they were often limited to only those hospitals with which either they or their DHA had contracts. So from '*talk*' about patient choice, the reforms descended through '*decisions*' which meant 'choice by proxy' into '*actions*' which actually limited choice more than before in many cases!

Brunsson himself suggests that 'there seems to be a good deal of hypocrisy at the societal level as well'. Major changes take place without discussion whilst areas of big change go unremarked (Brunsson 1989 p. 29).

Interestingly, Brunsson does not see hypocrisy within and by organisations as wholly dysfunctional. On the contrary, hypocrisy allows the organisation to tell 'stories' about itself which satisfy some of the competing demands placed upon it. They are able to create an external 'party line' about what the organisation is doing and how it is doing it which appears consistent and coherent, even when everyone inside the organisation knows that things are not really like that at all.

This is a problem which bedevils organisational and policy researchers by the way — going in from 'outside' one is invariably told the 'official' version of the organisation or policy which rarely tells the whole truth. I have done a great deal of work as a consultant and it is striking how much people inside the organisation are prepared to admit to problems, contradictions and gaps in what they are doing whilst presenting a supposedly coherent and rational version to the outside world. This is especially true in public organisations, where I have done most of my work, as they try to satisfy political and/or hierarchical masters that they are doing as required whilst they know they are not — or at least not as wholly or consistently as they suggest.

Hypocrisy can then provide a useful function — just like social 'white lies'. It is when, as Kolakowski points out about social hypocrisy, these deceits become self-deceits that they become dangerous (Kolakowski 1999).

Humans then aspire to consistent morality, to truth telling and to consistency and yet at all levels of our lives — from the closest inter-personal interactions, through casual acquaintanceships to organisational life — we indulge in hypocrisy. Sometimes this can be what might be termed 'noble cause' hypocrisy and it fulfils an important function in allowing us to live with paradoxes which would otherwise make life intolerable. Sometimes however it becomes deceit and self-deceit for far less noble motives and is highly dysfunctional. Living with paradoxical social instincts might be adaptively advantageous, but it is not easy.

Part II

EVOLUTIONS

In Part II we turn to the issue of how humans have ended up as paradoxical primates living in paradoxical societies with paradoxical institutions, cultures and organisations.

The explanations we seek to develop are evolutionary — that humans evolved from primates who already had some of the characteristics described, in terms of paradoxical instincts, but that we have developed and refined these through both natural and cultural evolution. This Part concentrates on the natural evolution of paradoxical social instincts in humans and tries to outline what they are, where they come from and finally some ideas about where we go from here in understanding them better.

Chapter Four

The Whisperings Within

I have no great sympathy for the Greek sage who asserts that war is the father of all things, but with a better right this honorific title might be given to conflict. Conflict between independent sources of impulses is able to produce, within the organism, tensions which lend firmness to the whole system, much as the stays of a mast give it stability by pulling in opposite directions. (Lorenz 1963 p. 80)

During much of the twentieth century the term 'instinct' was deeply unpopular in academic circles. This was because, as Pinker has persuasively argued, most of the social sciences were dominated by the 'blank slate' view of human nature (Pinker 2002). Pinker was also one of those who has helped in the rehabilitation of the term (Pinker 1994). What do we mean by 'instinct'? The term actually embraces two features of animal behaviour (including the human variety): a desire, motivation or drive to behave in a particular way and an ability to do so or learn how to do so. Thus humans are both driven to acquire language and have an ability to do so.

This Chapter draws on the proceeding evidence about paradoxes in human nature and other evidence from ethology and other sources to suggest a structure for paradoxical human instincts.

1. Hypothesising Paradoxical Instincts

Given the above definition of instincts and the previous discussion of paradoxes, what might paradoxical human instincts, which might explain some of the social, institutional and organisational paradoxes already explored, look like? We will restrict ourselves to some of the most obvious and usually hotly-contested aspects of 'human nature' in a social context: issues like whether we are innately aggressive or pacific; competitive or cooperative; conformist or autonomous; selfish or

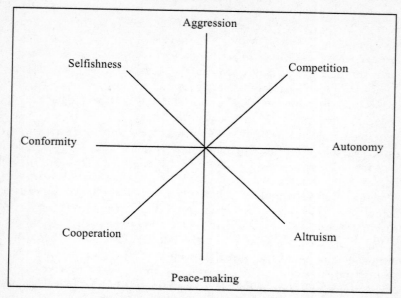

Figure 6: Paradoxical Human Instincts?

altruistic? These contradictory views are mapped in Figure 6, using the same approach as Quinn and colleagues to map managerial paradoxes (Chapter 1).

2. Aggression and Peacemaking

The 1960s and early 1970s saw a number of books published allegedly suggesting that humans are innately aggressive and that this comes from our evolutionary heritage. Amongst the most famous (or infamous depending on your point of view) of these texts were *The Imperial Animal* by the aptly named Lionel Tiger and Robin Fox (1971), the 'popular science' text *The Naked Ape* by Desmond Morris (1967), *The Territorial Imperative* by Robert Ardrey (1967), and perhaps the most important study, reviled and respected in equal measure, Konrad Lorenz's *On Aggression* (1963).

Published at the height of the Cold War, in the aftermath of the Cuban missile crisis, with opposition to the Vietnam War growing, and with the mass slaughter of World War II still a raw memory, these arguments for innate human aggression seemed inherently plausible to many. They seemed equally lamentable to those who blamed not evolution but human created institutions (e.g. capitalism, fascism, militarism, nationalism, religion, patriarchy) for our seemingly limitless capacity for violence.

Some anti-war groups even condemned these books or their authors as justifying American aggression in Vietnam — which was of course non-sense. Even if the books were as one-sided as alleged in their depiction of innate human aggression there were two sides at war in Vietnam, both displaying pretty aggressive tendencies, so such 'aggression is natural' views could presumably have been used to justify either side's belligerence, not just the Americans?

In this highly charged atmosphere these texts were also seized upon by different political interests as demonstrations of wider ideological positions — either the naturalness of capitalist individualism or the necessity of socialist collectivism. What rarely occurred, of course, was any measured debate of the relative merits of these studies and their actual message was often distorted, simplified and abused to serve specific interests (just like the work of Stanley Milgram in the next section).

Tiger and Fox, for example, were not simple instinctivists. They argued that humans were 'wired' like computers to process certain types of information but not pre-programmed — they were set to receive, process and transmit certain types of information at different points in their life cycle (the most obvious being language acquisition in early childhood). This applied to a wide variety of behavioural mechanisms. They differed from simple instinctivists by seeing individual human behaviour as broadly shaped by these information processing tendencies, the details of which would be supplied by the specific cultures in which individuals develop (Tiger and Fox 1971 p. 33). This obviously differs considerably from the blank slate approach, which they also specifically reject (p. 28).

So Tiger and Fox did not argue a simplistic view that humans are innately aggressive, but rather that we have a capacity for aggression which is shaped by our environment. Indeed, they quite rightly question the naïve assumption that 'aggression equals bad' whilst 'peace-making equals good'. Instead they argue that a limited form of aggression is a necessary part of any sexually reproductive species — for social animals this is better if it is non-violent and ritualised. However social animals like primates also need to be able to deploy a much higher degree of violence externally — either in self-defence or for hunting purposes. In their own words:

> That a species is characterized by aggression does not mean that its members do not cooperate and are never helpful or loyal or loving — quite the contrary. Aggression is an essential component of these virtues, and ethologists have argued very cogently that we should view it as a constructive motive, a positive force acting for the benefit of a species. Only when this positive force spills over into violence does it threaten internal order and survival. Even then, violence against predators can be seen as

another constructive device. There is a delicate balance to be achieved here; ethological studies have brought out brilliantly how various animal populations manage it. There must be sufficient aggression between members of the same species in the same population to ensure that selection takes place and various subfunctions are served; this must be contained — usually by the process of 'ritualization' — so that it does not develop into internal violence and destroy the population; violent opposition to threats from without the population must nevertheless be fostered. *Non-violent aggression within, and violent opposition on the outside, have constantly to be balanced in any human society: members must be capable of violence in order to preserve its integrity, but toward one another they should be non-violently competitive.* (Tiger and Fox 1971 pp. 250–1; emphasis added)

Perhaps the most widely read, and balanced, modern account of aggression and 'peace-making among primates' has been Frans de Waal's book of that name and his subsequent writings (de Waal 1989; 1996; 2001). De Waal's ideas have been seized upon by some to try to argue (like Rousseau) that humans are 'naturally' pacifists which is very far from de Waal's real views.

In detailed analyses of a range of primate species de Waal has shown that natural primate tendencies towards aggression are tempered by a wide variety of peace-making strategies. He argues that aggression is natural — but so is amelioration of aggression, especially in social species. Different primate species have different ways of dealing with aggression within the group — for example amongst single-male group golden monkeys the males invariably intervene to impose peace 'from above' between feuding females whereas amongst chimpanzees it is often females that use diplomatic skills to bring about reconciliation 'from below' between fighting males (de Waal 1996 pp. 31–3).

De Waal concludes that:

Our human societies are structured by the interplay between antagonism and attraction. Disappearance of the former is more than an unrealistic wish, it is a misguided one. No one would want to live in the sort of society that would result, as it would lack differentiation among individuals … If certain species, such as humans, reach a high degree of social differentiation, role division, and cooperation, this occurs because the cohesive tendency is counteracted by internal conflict. Individuals delineate their social positions in competition with others. We cannot have it both ways: a world in which each individual attains his or her own identity, and a world without clashing individual interests. (de Waal 1989 pp. 235–6)

He goes on to suggest that such combining of opposites — competition and cooperation; antagonistic and affectionate behaviour; etc. — produces very real 'paradoxes' which 'disturb the neat dichotomies that we set up to clarify our thinking'. He believes that such distinctions do exist, but that their 'intertwinement' in the long run makes the dichotomies appear false and prevents a proper understanding of the way 'conflict shapes our social life' (p. 237).

3. Conformity and Autonomy

There have been probably thousands of sociological and anthropological studies of under what circumstances humans conform to authority (individual or institutional) within their communities and when they rebel and assert their autonomy. However one study above all has come to symbolise the issue of conformity versus autonomy — the series of famous experiments reported in Stanley Milgram's *Obedience to Authority* (Milgram 1997). Few social scientists cannot have heard of Milgram's experiments. Even if they can't remember who conducted them or the exact details of the procedures, many have a hazy idea of what they were generally about and rather more firm ideas about what they 'proved' — people conform to authority to a rather frightening degree and are willing to do very horrid things when told to by their 'superiors'.

The experiments emerged from a question which exercised many after the Second World War — why had so many seemingly perfectly normal people (mostly Germans but plenty of other nationalities as well) participated in the programme of extermination directed against the Jews and others? Hannah Arendt wrote a book describing the trial of Adolph Eichmann, a leading Nazi who was captured and spirited to Israel by Nazi-hunters. The trial unveiled in huge detail the industrial nature of the Nazi extermination machine. Arendt coined the phrase the 'banality of evil' to describe Eichmann, a phrase which has now entered into popular consciousness as a synonym for blind obedience to evil authority (Arendt 1963).

Like the Nuremberg trials before it, Eichmann's trial raised yet again the issues of the complicity and culpability of thousands of ordinary Germans in the running of the death camps (and of others in occupied countries who collaborated in rounding up Jews). The oft-heard defence of 'I was just following orders' — which Eichmann and others employed — had no legal or moral justification. But it nevertheless appeared that many Germans did seem to have done just that: unquestioningly conformed to authority regardless of the horrendous implications of what they did (for those not directly involved in the slaughter) or even nature of the acts they were being asked to commit (for those who were).

Stanley Milgram and his social psychology colleagues were motivated by these issues to try to find out just how 'normal' it was for people to conform to instructions which were clearly immoral and/or illegal and involved inflicting harm on others. Their series of experiments (there were many variations), conducted in the 1960s, all revolved around a deceptive setup in which an experimental subject thought, wrongly, that they were actually participating as an assistant in another experiment. A supposed 'victim' (called in the experiments the 'learner') was seated and strapped into in a chair in another room, separated by a glass partition from the real subject (called 'the teacher'). The 'learner' had electrodes attached to him. The subject/teacher was to administer a test and whenever the learner got a question wrong, to administer an electric shock using an impressive looking 'shock generator'. The learner was actually an actor and the 'shock generator' a fake, but the 'teacher' did not know this. An experimenter (the authority figure) was also present to make sure the 'teacher' carried out the procedure correctly. The 'experimenter' also instructed the 'teacher', telling him at various points to raise the level of electric shocks being administered, even up to and above a clearly marked 'danger' level on the 'shock generator'.

Many people will recall the basic outlines of these experiments and will assert that what they showed was that most people were perfectly happy to obey the 'authority' and administer intense levels of shock, which provoked supposed cries of agony and pleading to stop from the 'learner' victim.

I and many like-minded colleagues have in the past, during arguments about nature versus nurture, cited these experiments as evidence of how much humans are willing to conform to institutions, and argued that 'nurture', or in this case the cultural/institutional environment, can always be seen to triumph over any innate or intrinsic values or views of the individual. Everybody conformed, with only a very few exceptions whose unique heroism simply highlighted how much everyone else simply went along with what was happening (one thinks of *Schindler's List*).

These experiments seemingly strong support to those who saw the institutional and cultural context of individuals as determining their behaviour explained 'wicked' behaviour in 'evil' social systems like Nazism. But it also offered the hopeful prospect that in 'good' social systems, like socialism, people would behave differently. If the institutions or culture are to blame, simply change them and people will change.[1]

---

[1] It is more than somewhat ironic that at its most extreme this 'environmentalist' view led supporters of the Maoist Cultural Revolution in China and the appalling 'Year Zero' in Cambodia to argue they were changing humanity by changing the system. In both cases the attempt was, supposedly, to create a 'pure' socialist environment by purging society of 'bourgeois' elements. In Cambodia's case this included exterminating anyone exhibiting 'bourgeois' characteristics in the 'Killing Fields'. It is thus somewhat ironic that today's defenders of environmental determinism choose to ignore their own bastard off-spring, in the Cultural Revolution and Year Zero, whilst happily trying to pin the blame for Nazi extermination camps on anyone who

Unfortunately, as usual with these things, the popular 'reading' of Milgram's experiments is deeply flawed and distorted. In fact, a considerable number of his 'teachers' rebelled against the instructions — they did not always conform and substantial numbers, sometimes even a majority, rebelled. The numbers of rebels varied with some significant factors like their distance from the 'victims' or the amount of peer as well as authority pressure exerted upon them. It was clear there were actually strongly contradictory impulses at work in most of the 'teachers', even those who complied, as a reading of the exchanges, questioning and worries they exhibited shows.

If it is possible to draw a conclusion from Milgram's experiments then it is that there were contradictory impulses and that these manifested themselves in contradictory behaviours, with the impulse to conform being overall probably stronger than the impulse to rebel, but both appear present for most people most of the time.

It is worth listening to what Milgram himself had to say on the subject, because his own interpretation is very different from that which has been most usually ascribed to his work:

> Let us begin our analysis by noting that men are not solitary but function within hierarchical structures. In birds, amphibians, and mammals we find dominance structure, . . . and in human beings, structures of authority mediated by symbols rather than direct contests of physical strength. The formation of hierarchically organised groupings lends enormous advantage to those so organised in coping with dangers of the physical environment, threats posed by competing species, and potential disruption from within. The advantage of a disciplined militia over a tumultuous crowd lies precisely in the organised, coordinated capacity of the military unit brought into play against individuals acting without direction or structure.
>
> An evolutionary bias is implied in this viewpoint; behaviour like any other of man's characteristics, has through successive generations been shaped by the requirements of survival. Behaviours that did not enhance the chances of survival were successively bred out of the organism because they led to the eventual extinction of the groups that displayed them. A tribe in which some of the members were warriors, while others took care of children and still others were hunters, had an enormous advantage over one in which no division of labour occurred. We look around at the civilisations men have built, and realise that only directed, concerted action could have raised the pyramids, formed the societies of Greece, and lifted man from a pitiable creature struggling for survival to

---

mentions biology and human behaviour in the same breath. Both sets of extremisms are deadly distortions of perfectly respectable scientific positions and it is about time we got away from such sterile 'guilt by association' rhetoric. Does anyone seriously think that the ideas of Steven Rose and the late Stephen J. Gould are responsible for socialist death camps any more than they think the ideas of E.O. Wilson and Stephen Pinker condone Nazi ones? I sincerely hope not.

technical mastery of the planet.

The advantages of social organisation reach not only outward toward external goals, but inward as well, giving stability and harmony to the relations among group members. By clearly defining the status of each member, it reduces friction to a minimum. When a wolf pack brings down its prey, for example, the dominant wolf enjoys first privileges, followed by the next dominant one, and so on down the line. Each member's acknowledgment of his place in the hierarchy stabilises the pack. The same is true of human groups: internal harmony is ensured when all members accept the status assigned to them. Challenges to the hierarchy, on the other hand, often provoke violence. Thus, a stable social organisation both enhances the group's ability to deal with its environment and by its relationships reduces internal violence.

A potential for obedience is the prerequisite of such social organisation, and because organisation has enormous survival value for any species, such a capacity was bred into the organism through the extended operation of evolutionary processes. I do not intend this as the end point of my argument, but only the beginning, for we will have gotten nowhere if all we can say is that men obey because they have an instinct for it. Indeed, the idea of a simple instinct for obedience is not what is now proposed. Rather, we are born with a potential for obedience, which then interacts with the influence of society to produce the obedient man. In this sense, the capacity for obedience is like the capacity for language: certain highly specific mental structures must be present if the organism is to have potential for language, but exposure to a social milieu is needed to create a speaking man. In explaining the causes of obedience, we need to look both at the inborn structures and at the social influences impinging after birth. The proportion of influence exerted by each is a moot point. From the standpoint of evolutionary survival, all that matters is that we end up with organisms that can function in hierarchies.' (Milgram 1997 pp. 141–143)

When Milgram notes that a substantial proportion of his subjects at some point rebelled against the orders to 'shock' their victims, he ascribes this to a more fundamental design compromise in human beings, related to our ability to function *both* individually *and* collectively:

Theoretically, strain is likely to arise whenever an entity that can function autonomously is brought into a hierarchy, because the design requirements of an autonomous unit are quite different from those of a component specifically and uniquely designed for systemic functioning. Men can function on their own or, through the assumption of roles, merge into larger systems. But the very fact of dual capacities requires a design compromise. We are not perfectly tailored for complete autonomy, nor for total submission. (Milgram 1997 p. 171)

In other words our nature as both autonomous beings and social beings generates fundamental tensions which have to be constantly resolved. We are not entirely unique in this regard — some other social beings (other primates, some pack-hunter animals like wolves) can associate and even to a limited extent cooperate in groups or act individually as the circumstances demand. However the degree of human ability to function within a group and conform to instructions and at the same time to be able to function completely independently is remarkable and probably of a higher order than any other species (we know about). Of course language and other cultural artefacts play a key role here in being able to transmit information and, very importantly, reinforce the expression of particular instinctual biases in particular directions — what sociobiologists have called 'gene-culture co-evolution' (Wilson 1975). We will return to this subject later. For now, let us turn to another paradoxical pair — altruism and selfishness.

4. Altruism and Selfishness

The dominant social science of the twentieth century was undoubtedly economics. Not only has this been the pre-eminent discipline but it has also invaded the territory of other social sciences — for example through the incursions made by so-called 'public choice' theory into politics and public administration (e.g. see Dunleavy 1991; Self 1993).

At the core of neo-classical economics, the dominant approach within the dominant social science, is the notion of human beings as 'rational utility maximisers' — that humans have a set of fixed preferences which they seek to obtain through the least cost to themselves by a process of rational calculation of the costs and benefits of any particular course of action.

Let us just pause for a moment before we start examining this idea in more detail and consider what sort of theory this is. It is essentially an evolutionary theory — something which most economists will hotly dispute. Note that the assumptions do not allow for variation due to culture or institutions — something indeed over which there has been intense and often acrimonious debate amongst economists but with the 'orthodox' tendency strongly defending the position that all individuals everywhere are the same. Now, consider how all individuals everywhere, regardless of cultural and institutional environments, could be exactly the same in their 'economic' behaviour? If these behaviours were learned they would be subject to the vagaries of both individual experience and cultural and institutional environments. They must therefore be innate. To be innate they must be inherited and to be inherited traits humans must have evolved to be that way. Any 'human universal' (Brown 1991) must, by definition, be innate and inherited rather than learned. How could it be anything else and still be 'universal'? Any other explanation would admit relativism in human behaviour which would fundamentally undermine the 'rational utility maximiser' hypothesis.

In this respect at least economics has run counter to the 'blank slate' (Pinker 2002) or 'Standard Social Science Model'[2] (Barkow, Tooby et al. 1995) which has been dominant in the rest of the social sciences. Whilst most neo-classical economists would deny absolutely any necessary link to a Darwinian explanation of human behaviour, their core theoretical beliefs cannot be explained any other way.

Neo-classical economic assumptions are hardly uncontroversial how-ever — both from within economics and from without they have been subject to extensive criticism. We will examine those criticisms which are most relevant to the thesis of the Paradoxical Primate we are developing here. The 'blank slate' critique is not of much relevance, but some other critiques are — including institutionalism and critiques which suggest innate contradictory preferences (especially altruism) and non-rationality in human decision-making.

This chapter turns to evidence from economics about paradoxical human behaviour and motivation.

Contradictory Preferences

The first critique of neo-classical economic assumptions about human behaviour which we'll look at is that which suggest that the concept of 'utility' is far too simplistic and assumes far too much consistency in human preferences.

The most obvious criticism here is the rather obvious one of altruistic acts. Humans throughout recorded history, in all societies, have engaged to a greater or lesser degree — sometimes even a heroic degree — in acts of pure altruism. This is so obvious that we need hardly bother marshalling any great body of evidence. How does neo-classical eco-nomics respond? Mostly by either ignoring the problem or by redefining personal 'utility' in so broad a way that it ends up encompassing any-thing anyone could ever do. 'Altruism' is — in this response — redefined as selfishness, i.e. the individual committing an altruistic act is in fact gratifying their own preferences for feeling good about themselves by being altruistic. (Now all you have to do is redefine 'up' as 'down' and you have a complete scientific explanation of the universe!) Aside from its obvious flaws (how exactly is dying to protect someone else going to allow you to feel good about yourself?) this little dodge actually under-mines what explanatory power economics does possess.

2]   In this one respect the SSSM is clearly not a completely 'standard' model — it clearly does not apply to neo-classical economics which does make (implicit) assumptions about human nature.

There is another more intriguing way of looking at this defence of neo-classical economic assumptions. According to this argument a single human can have a preference for *both* selfish egotistical acts *and* for apparently unselfish altruistic acts — which perhaps sounds rather paradoxical? Of course its proponents would suggest that they are entirely rationally consistent — both selfish and altruistic acts are actually both selfish so there is no paradox or inconsistency. Others beg to differ. We turn to look at some alternative explanations which accept (albeit usually not in the same terms we are using) the paradoxical nature of human preferences. What follows are two examples: one from economics and one from public policy, both of which explore — if not explicitly — paradoxical human motivation.

G-Smith and S-Smith

Howard Margolis is professor of public policy in the heartland of neo-classical economics — the University of Chicago. This did not stop him coming up with a radical challenge to the simplistic notion of rational utility maximisation in a theory of social choice (Margolis 1982).

Social choice was developed in clear opposition to, and as a solution to some of the contradictions and paradoxes within, the so-called 'public choice' school of economics which sought to apply neo-classical methods to political problems and for which Chicago was most famous.

Margolis points out a number of paradoxes such as the so-called 'paradox of voting' — why should any individual bother to vote as the cost of voting far outweighs any possible individual benefit due to the minute chance that their personal vote makes any difference. Similar paradoxes, according to Margolis, include the 'prisoner's dilemma' and the 'paradox of public goods':

> Suppose we consider some everyday approximation of a pure public good that is supported by voluntary contributions: for example, a listener supported radio station. Empirically, there is no question that the amount of money that can be raised from listeners will depend strongly on the number of listeners. The paradox is that it can be 'proved' (in terms of the usual model of rational choice) that the amount of money that can be raised will be very nearly independent of the number of listeners: the station should be able to raise about as much money if it had five listeners as if it had 50,000. Or, to put the same paradox another way, because a station could not raise significantly more money from 50,000 listeners than it could from 5, and because 5 listeners would not contribute enough to keep the station on the station on the air, then we deduce that listener-supported radio stations cannot exist. But they do. (Margolis 1982 p. 19)

As an interesting aside, Margolis adds a telling remark about the state of economics:

It is an interesting comment on the state of the field that when journal articles appeared carrying this implication — one by a political scientist, another by an economist — neither the authors nor, apparently, anyone else was moved to remark that there was something bizarre about such results. (ibid p. 19)

As an alternative Margolis develops a theory of what he calls the 'fair-share' (FS) model which is based on individuals having two separate sets of preferences — what he personifies as 'S-Smith' (self-preferring Smith) and 'G-Smith' (group-preferring Smith). Actual Smith must choose between allocating fixed resources between his or her 'S' preferences and 'G' preferences. Margolis goes on to develop detailed analyses of how this process might unfold in various situations with all the usual mathematical explanations and diagrams of economics. But his approach can be summed up in the simple rule he postulates:

The larger the share of my resources I have spent unselfishly, the more weight I give to my selfish interests in allocating marginal resources. On the other hand, the larger benefit I can confer on the group compared with the benefit from spending marginal resources on myself, the more I will tend to act unselfishly. (ibid p. 36)

This rule suggests the sort of elastic tension between paradoxical elements that have already been encountered in these pages. The more Smith veers towards 'S' preferences, the more likely he/she is to prefer 'G' preferences in marginal decisions and *vice versa*.

Margolis also offers an explicitly Darwinian explanation of how such a set of dual preferences could evolve which is worth quoting in full:

An immediate consequence of the Darwinian analysis is that self interested creatures, other things being equal, will be able to leave more descendants carrying their genes than would non-self-interested creatures. Hence natural selection will favour self-interest. (As always in a Darwinian context, 'self-interest' must be understood to include kin-altruism, where the helping behaviour yields direct advantage in terms of inclusive fitness. This usage is not, in fact, very far from that ordinarily seen in economics.)

On the other hand, in competition among breeding groups, within some of which individuals are endowed with a propensity to act in the interest of the group (rather than solely in self-interest), the groups whose members have a propensity to act in group-interest will have a selection advantage over the groups deficient in that propensity.

So we can imagine another kind of competition, between self-interest and group-interest, as the propensity that will be most favoured by natural selection. Within a group, self-interest will be favoured; among groups, group-interest. The possibility then arises that the outcome might be a compromise of some kind, especially in species in which there is great potential advantage to groups that are able to sustain a high level of cooperative behaviour beyond the level of small kin-groups. In particular, much human behaviour seems to involve some such compromise. (Margolis 1982 p. 26)

We will return to the issue of group selection later.

Of Knights and Knaves

Julian Le Grand is a professor of social policy at the LSE (London School of Economics) and also, at the time of writing, an adviser in the Prime Minister's Office in 10 Downing Street. A book he published shortly before entering Number 10 postulates a view of human behaviour in public organisations which is remarkably similar to Margolis's analysis (Le Grand 2003).

Le Grand coins the very English terms 'Knights' and 'Knaves' to describe the equivalent of 'G' Smith and 'S' Smith. Whereas Margolis looks at general human behaviour, Le Grand concentrates on people working in public service and how far their behaviour is motivated by altruistic and selfish desires. Le Grand argues that public policy makers tend to make assumptions about people being either knights or knaves:

> Assumptions concerning human motivation — the internal desires or preferences that incite action — and agency — the capacity to undertake that action — are key to both the design and the implementation of public policy. Policy-makers fashion their policies on the assumption that both those who implement the policies and those who are expected to benefit from them will behave in certain ways, and that they will do so because they have certain kinds of motivation and certain levels of agency. Sometimes the assumptions concerning motivation and agency are explicit; more often they are implicit, reflecting the unconscious values or unarticulated beliefs of the policy-makers concerned.
>
> Conscious or not, the assumptions will determine the way that public polices are constructed. So, for instance, a policy instrument designed on the assumption that people are motivated primarily by their own self-interest — that they are, in the words of David Hume quoted at the beginning of this book, *knaves* — would be quite different from one constructed on the assumption that people are predominantly public-spirited or altruistic: that they are what we might term, in contrast to knaves, *knights*. (ibid p. 2)

Public policy has in fact tended to oscillate between these two sets of assumptions, according to Le Grand, whereas in fact the research evidence tends to suggest that people are not either knaves or knights but both at different times and in different circumstances. He coins the term 'act relevant knights' to depict the altruistic aspect of behaviour and explores research evidence that people can indeed by entirely altruistic — as well as selfish — but within certain limits. Indeed he suggests that altruistic acts — at least in a public service context — follow a rather counterintuitive pattern:

The relationship between the benefit derived from an altruistic act and its opportunity cost will be complex. If the cost is too little, the benefit from making the sacrifice will also be relatively little and the individuals motivation to perform the act will be correspondingly reduced. For the activity to feel really worthwhile, people need to feel that they have made some effort to perform it to have incurred some significant cost. Too great a cost, on the other hand, will also demotivate them; they will feel that there is a limit to the amount of sacrifice of their interests that they are prepared to make for the sake of others, and they will therefore feel less inclined to undertake the activity. In other words, there are cost thresholds such that, if the cost falls below the lower threshold or rises above the higher one, people are less likely to perform the activity than if the cost falls in between. (Le Grand 2003 p. 52)

What both Margolis and Le Grand argue is that humans have both selfish and altruistic (group oriented) motivations. Margolis makes this an explicitly evolved trait, whereas Le Grand (in common with most neo-classical economists) just assumes this to be a universal human condition without making the obvious link to a genetic basis, or rather genetic and cultural origin, for such behaviour (and its variations).

5. Cooperation and Competition

As Josiah Tucker pointed out in 1781 (cited above), humans are naturally cooperative animals. Throughout what we know of human history groups of us have banded together and, perhaps more importantly, worked together through some form, however rudimentary, of a division of labour.

One thing [scientists] do not disagree about, however, is whether humans are social creatures. We are, period. We know of no cases throughout history where large numbers of humans have intentionally lived outside the fabric of *some kind* of society. (Dugatkin 2000 p. 2; emphasis in original)

Of course, individuals have chosen a solitary life — various hermits, explorers and others. Such examples are the only real challenge to a view of humans as inherently sociable — they are often used by proponents of the 'blank slate' view of human nature as examples of how humans are formed by culture and choice rather than genes. However, we would not use the few isolated examples of fish that do not migrate to their spawning grounds or birds that fail to fly south for winter to suggest that these are not migratory species. So why should a few hermits cause us to dismiss the idea that we are an inherently sociable and cooperative species?

Sociability is a lot more than mere association. Some species do tend to group together but where there is no or little exchange of resources or work between these flocking or herding creatures. They do often gain some benefits from association — the most usually mentioned is the protection against predators. Within such herds some elementary forms of cooperation do begin to emerge — for example the creation of 'nurseries' where parents (usually mothers) protect offspring in groups. But apart from protection and such elementary forms of cooperation, such herds are hardly societies.

Social organisation — which we might define in terms of some form of division of labour and sharing of resources — has arisen again and again in a wide variety of species. Invertebrates, insects, and mammals have all produced various forms of social organisation — like flying, social organisation has evolved very frequently although in rather different ways.

Cooperation, it should be clear, is not the same as altruism or even conformity. Pure altruism implies the giving of something without any anticipation of reciprocity. Cooperation on the other hand assumes some form of reciprocal exchange — although confusingly the term 'reciprocal altruism' is often used (following a seminal paper by Robert Trivers [1972]). Similarly, one can conform without necessarily cooperating. Cooperation implies an active and willing exchange of resources in some mutually beneficial way.

Behavioural ecology has come up with four broad sets of reasons why any animals cooperate: kinship; reciprocal exchange; selfish teamwork; and (most controversially) group selection (Dugatkin 2000 pp. 17–24).

Kin support is the most obvious evolved form of cooperation — protecting or supporting genetically close relatives increases the possibilities of 'your' genes, or ones very much like them, surviving and reproducing. Kin cooperation is thus only a small extension of 'cooperating' to one's own off-spring to make sure they survive. However, kin-based cooperation does have obvious draw-backs such as the relatively rapid drop-off in genetic identity (siblings have the same parents; cousins only the same grand-parents; and so on). How to identify kin as opposed to others of the same species can also be problematic. Nevertheless kin cooperation does adequately explain some cooperative behaviour, such as the alarm-call dilemma — whoever makes the alarm call about an approaching predator makes themselves a more likely target for said predator. Despite this rather obvious risk, individuals in a wide range of species do make alarm calls (Dugatkin 2000; Winston 2002).

Reciprocal exchange, or Tit-for-Tat as it is often known, is based on the idea of cooperation to achieve a mutually beneficial outcome for pairs of individuals. Most famously this strategy was proved by Robert Axelrod through the use of a computerised game of 'Prisoner's Dilemma' in which the player's strategy was to reward cooperation and punish defection — so-called Tit-for-Tat. Over a number of iterations cooperation emerged as the dominant mode of behaviour (Axelrod 1990). The Tit-for-Tat solution relies crucially on a couple of factors — individuals must regularly be involved in similar choices and the long-term benefits of cooperation must outweigh the short-term benefits of defection. It also suffers from the problem that as soon as larger numbers of 'players' — n-person games — are examined the Tit-for-Tat strategy breaks down as players have no way of focussing punishment easily on defectors. As Robert Axelrod points out, despite this difficulty cooperation clearly does emerge. Further modelling work led Axelrod to the conclusion that there were ways in which norms could emerge and be sustained which promoted cooperation, although the resulting models start to be a bit more complex (Axelrod 1997 p. 7 and Chapter 3).

Selfish teamwork is where cooperation is also mutually beneficial but because the cooperators achieve a result which is simply not possible without teamwork. In Tit-for-Tat situations individuals can gain short-term benefits from acting alone — but such situations are very specific (Prisoner's Dilemma is a very specific set of circumstances). Much more usual are situations where individuals simply cannot achieve as much alone. Many hunting species — for example lions and dolphins — rely on teamwork to achieve results which would be difficult if not impossible for individuals acting alone. As long as the collective effort can produce results which benefit all the team members and only the team members then here we have selfish cooperation. This is the type of cooperation much admired by Adam Smith and celebrated as the 'invisible hand' of the market which produces coordinated action from selfishly motivated individuals.

The fourth cooperative strategy is group altruism — that is where individuals sacrifice for the good of the group, who need not necessarily be close kin or even related at all. We don't have to search far for many, many examples of where individual humans engage in acts of group altruism — it is so self-evident it hardly needs support. The real issue, from an evolutionary perspective, is not does it occur, but why does it happen?

Cooperation based on group altruism requires group selection. The idea of group selection is still a highly controversial one in evolutionary biology. It has been condemned by the majority, but there has recently been something of a resurgence in interest in the idea and substantial theoretical and empirical support (Sober and Wilson 1998). The key to the possibility of group selection is the idea of multi-level selection — i.e. that what is good (or bad) for replicating a particular gene may be different at different levels of biological activity — genome, cell, organism, group and species. A good negative example are cancer cells — they are very good at replicating themselves in the short term but in the long term they destroy their hosts. (Many viruses have similar traits.)

Elliot Sober and David Sloan Wilson have perhaps been the most prominent recent exponents of group selection (to explain altruism) and have developed a sophisticated analytical framework for establishing selection pressures at different levels of biological activity and then combining these to see what actually happens. They point to some interesting statistical phenomena involved, such as 'Simpson's Paradox'. This shows that calculations showing a decrease in altruists in all groups of individuals can reverse at the species level if the result of having altruists in some groups is that they reproduce faster than groups with lower proportions. In all groups the proportion of altruists decreases, but overall their numbers increase, because of group selection effects (Sober and Wilson 1998 pp. 23–6). In their analysis it is crucial that altruists are altruistic towards the group — in other words they are group co-operators. It is this which produces differential outcomes for groups, and the altruistic co-operators within them.

Group altruism is closely related to the 'reciprocal altruism' idea developed by Robert Trivers (1972). The reason Trivers called his theory 'altruistic' was because of the time lag between acts of giving and acts of receiving — he suggested that these acts were not directly associated and therefore had to be considered 'altruistic' even when they were later reciprocated. Reciprocal altruism clearly only works for individuals who meet regularly.

There is a fifth idea about cooperation — what has been called the Big Mistake hypothesis. Like the peacock's tail, something which evolved for one purpose (sexual selection) can end up being a benefit or hindrance in other respects. In the case of cooperation the idea is that kin-based cooperation which evolved within species that live almost entirely in small groups may have been just a general inclination towards cooperation, which benefited all members of the group to whom the co-operator was usually related. However, once the species evolved into different living patterns — mixing freely with non-kin groups for example — the general instinct towards cooperation got transferred to non-kin simply because it was there. In this argument, humans who evolved in small groups developed what was initially a kin-based cooperative instinct but as they progressed to form large groups and communities the cooperative instinct remained (Winston 2002 pp. 313–15).

All five of the above explanations of the evolution of the human instinct for cooperation might be at least partially true — they are not necessarily mutually exclusive. What is without question is that the instinct for cooperation is a 'human universal' (Brown 1991). But we also compete — between individuals, family groups, groups, teams, clans, nations, societies.

Competition amongst humans (and between humans and other species) is hardly surprising — the dominant form of interaction between and within species on our planet is competition. Indeed, that is why explaining the origin of cooperation has proved so difficult from an evolutionary perspective.

What is really interesting is that only amongst humans and a few other species has there evolved such a strong and flexible capacity for both cooperation and competition. Most social species which have evolved cooperative traits are fairly rigid about their application — for example symbiotic species, invertebrate and insect colonies — whereas humans, other primates and a few other relatively complex species (such as cetaceans) seem to have evolved a special cooperative–competitive flexibility.

6. Conclusion

Humans, it has been argued in this Chapter, have a set of evolved paradoxical instincts. Are we alone in such contradictory predilections? Hardly — there are many other primate species and indeed other of the more intelligent animals that have at least some, if not all, of these tendencies. Indeed having contradictory instincts would seem to be a necessary basis for evolving complex and therefore adaptable behaviours. The reason it is so important to emphasise their appearance in human heritage is to show that some of our Western, rationalist, assumptions about human behaviour which tend to exclude paradoxical behaviour are wrong.

This is not to suggest, it has to be quickly added, that only contradictory or paradoxical explanations are sufficient to explain human behaviours and institutions. There are also features which are not contradictory and are more amendable to traditional, rationalist, modes of analysis. The crucial failing in a lot of social science is to conflate rationalist with rationalising explanations — ignoring the real contradictoriness of human behaviour (or ascribing it to purely relativistic sources as in social constructionist explanations).

Chapter Five

The Descent of Man

This Chapter[1] addresses the issue of how humans might have evolved into the paradoxical primates which are suggested in this book. I start by looking at some models of mental evolution which perhaps point to ways in which paradoxical instincts might be tied into our models of how human cognitive capacity evolved.

Next I look at how paradoxical instincts may have helped individuals to become more adaptable — a feature of many higher species but especially prominent in humans and which has bedevilled social science research for decades.

Humans are not merely individuals however, we are also very social animals — but of a particular type — the paradoxical instincts discussed throughout this book are above all social instincts directed towards how we live with our fellow humans. A small thought experiment will be conducted about how paradoxical primates might fare compared with either rather less social humans (a group of 'Yobs') or rather more social humans (a band of 'Hippies').

The idea of 'fission-fusion' societies has developed only quite recently in ethology but it has potentially very strong implications as to how not only paradoxical instincts might have evolved but crucially how they function in a social context.

1. Instincts, Emotions and Intellects

How do the paradoxical instincts explored in Chapter 4 interact with other individual human faculties — particularly emotion and intelligence? There is now a vast literature on how human minds might have evolved and others are far more expert in this field. I am therefore not proposing a theory here of how human minds may have evolved, I am merely 'cherry-picking' some ideas which might relate to my specific thesis about a paradoxical component to human behavioural patterns and hopefully be suggestive of lines of further enquiry.

[1] Apologies for the sexism in the heading, but that was Darwin's title.

Remember that up until now what has been examined are instincts — a predisposition towards, and faculty for, a certain type of behaviour. That such behaviours — sometimes quite complex in nature — can be inherited is not in doubt, as there are numerous animal examples we could draw upon to illustrate this.

Probably one of the most influential recent books on the role of emotion in humans has been LeDoux's *The Emotional Brain* (1998). This work is also based on the evolved structuring of the brain and emphasises the role of the limbic system and its interactions with, and to some extent shaping of, the more rational higher cognitive functions. This is a subject we will return to later. For the moment it is worth noting that LeDoux (and many others) have concluded that the brain evolved an emotional system that is both instinctual — that is partially 'hard-wired' — and capable of learning. Moreover this emotional system is actually not one system but several inter-related systems. LeDoux suggests that emotions are rather like sensory functions in the sense that all senses are sensory functions but they have separate neural structures underpinning sight, smell, hearing, etc. Likewise, he suggests emotional functions have separate neural structures and processes associated with each of them that have evolved over time.

The most popular model of these multiple emotional systems is Plutchik's 8 Basic Emotions (see Figure 7). This is not the only categorisa-

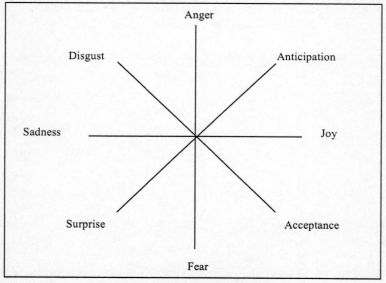

Figure 7: Plutchik's 8 Basic Emotions
(Source: Adapted from LeDoux 1998 p. 114)

tion available but LeDoux and others have recognised this as being relatively robust. The Plutchik model suggests only 8 basic emotions but that these can combine to produce higher level emotional states. Thus, for example, adjacent emotions such as joy and acceptance can combine into a state of friendliness; or fear plus surprise into alarm.

Perceptive readers will have already noticed however that some of Plutchik's basic emotions seem to be not complementary and synthesizable but more antagonistic and contradictory and this is emphasised in the layout in the diagram above (which is remarkably similar in configuration to many other attempts to map paradoxes). These opposites produce the most extreme emotional states — e.g. fear plus anger can produce rage, or disgust plus acceptance can produce shame or humiliation. But they can also produce emotional states we refer to as 'conflicting emotions' which we find especially hard to cope with. Thus gaining a promotion (joy) at the expense of your best friend (sadness) can cause extremely unstable emotional reactions.

We should be clear that we are not suggesting here that paradoxical human behaviour stems from paradoxical emotional traits. As LeDoux elegantly shows the expression of emotions in humans (and indeed the self-consciousness of emotional states) involves a complex set of interactions including both the basic emotions and higher cognitive functions. Nevertheless, at the very least we can suggest that the existence of potentially conflicting emotions can be associated with, and reinforce, conflicting values and instincts for paradoxical behaviour in humans.

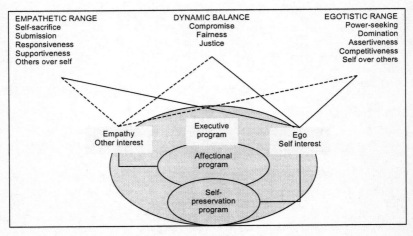

Figure 8: CSN Model

(Source: Adapted from Cory in Somit and Peterson 2003)

A model which comes very close to the 'paradox hypothesis' being developed in this book is the Conflict Systems Neurobehavioral (CSN) model (see Cory in Somit and Peterson 2003). This is a development on MacLean's 'triune' model of the brain which suggests the human brain has evolved by adding successive layers of structure — the reptilian complex, paleomammalian complex and finally the neomammalian complex (the neocortex).

This model clearly suggests a basic conflict between self and other (group) oriented programmes. In order to manage the fundamental conflict between these opposed systems a third 'executive programme' is required which chooses when to express each approach. The model clearly suggests a constant competition between the two systems — ego and empathy — which never really gets resolved, apart from temporarily tilting in one direction or the other.

The two fundamentally conflicting systems — ego and empathy — sound remarkably like the 'S-Smith and G-Smith' or 'Knaves and Knights' explored previously, and indeed like the whole conflicting model of human instincts outlined in Chapter 4. Whilst the list of competing characteristics — self-sacrifice, submission, responsiveness, supportiveness, and others over self *versus* power-seeking, domination, assertiveness, competitiveness and self over others — are not cast in exactly the same language as the competing instincts model set out in Figure 6, the gist of the characteristics are remarkably similar. Whether the CSN model or mine is correct in detail or not, something like this surely captures the essence of what is being argued in this book? We humans are essentially conflicted between our individual and social selves and a great deal of our behaviour derives from this basic paradox.

The CSN Model emphasises the need for an 'executive programme' which constantly adjusts the contending forces of ego and empathy. The role of intelligence is therefore crucial in choosing which set of instinctive responses to deploy at any particular moment. Intelligence is of course not just about computation at a particular moment, based on current inputs, but it is also about memory and learning. Learning which of the conflicting 'whisperings within' to listen to in which circumstances is surely part of the developmental process for all humans (and for that matter many other primates).

There are then emerging models of how both emotion and intellect might relate to paradoxical human instincts. This is a new field of exploration and the 'paradoxical' element is not yet widely recognised amongst those engaged in evolutionary psychology and behavioural development studies — it has thus far been mainly restricted to organisation and management studies, a field they are unlikely to be familiar

with. The independent emergence of approaches which are similar to the
'paradox' school in organisation and management within these fields is
therefore all the more striking.

In the next two sections the issue of the benefits to individual and
group adaptability of paradoxical instincts will be briefly explored.

2. Paradoxical Instincts and Individual Adaptability

There seems to be an obvious but not often remarked correlation in the
development of species — the more biologically complex they become
the more behaviourally adaptable they also become. Behavioural adaptability should not be confused with biological adaptability — indeed in
some ways the more biologically and behaviourally complex species
become they seem to (mostly) become less environmentally adaptable.
In other words, they become more flexible in a narrower range of potential habitats. (Humans seem to have broken this rule by finding artificial
ways of carrying our preferred environments around with us). Behavioural flexibility and adaptability has evolved way beyond the simple
autonomic responses of simpler creatures.

An example — John Alcock tells a wonderful story about a study that
he and a colleague made of the little rove beetle, *Leistotrophus versicolor*
(Alcock 2003 pp. 154–60). The male of the species, they observed, has a
rather curious variable reaction to other males when trying to occupy a
territory and acquire females and food. If confronted with a smaller male
they tend to attack and try and chase it away. But if they are the smaller
male sometimes they, instead of fleeing, start to imitate the motions of a
female rove beetle, teasing the larger male into following them and carrying out courtship rituals. They have even been observed to manage to
mate with females whilst the larger male waits patiently for an opportunity to carry on courting them!

This strange behaviour is not, according to Alcock, due to any detect-
able genetic variation between individual rove beetles — it seems all
males can adopt this 'conditional strategy'. So confronted with an environmental stimulus — another male — the rove beetle can either attack,
flee, or imitate a female. In other words it has a pretty flexible, adaptable,
response based on a repertoire of behaviours which are contradictory —
a paradoxical beetle. The rove beetle is a pretty simple soul, with little
'mind' to speak of, but it can deploy a variable set of behaviours in
response to graded versions of the same external stimulus — another
male.

This example shows precisely how as beings have gradually become more complex they have evolved more and more complex 'conditional strategies' rather than simple autonomic ones.

An organism which has only one possible reaction to a stimulus — a sunflower with the reaction 'move flower head to follow the light' — is clearly pretty unadaptive. (I don't know if this has been tried but sunflowers get their name from the fact that their flower-heads follow the sun across the sky everyday. So, if you put a sunflower in a light-proof room and get a daylight generating bulb to circle around it at approximately the right speed for the sun — would it eventually twist its own head off? If it does, that would be good example of how automatic responses to environmental stimuli can be inflexible and maladaptive.)

The rove beetle can adapt its behaviour to quite small changes in the nature of the stimulus and — like all more complex beings — precisely which behavioural response will be deployed is not deterministically predictable, only probabilistically. Not all smaller rove beetles, or even the same rove beetle in all cases, always react to larger males by pretending to be a female. Their responses are variable, if following a usual pattern.

Similarly, as soon as more than one goal comes into play an organism not only has to becomes more adaptive — having more than one response routine — but it also has to start to have a mechanism for choosing which goal to prioritise in which circumstances (otherwise it would simply freeze into immobility when confronted with choices between goals). Should it move towards the food or the warmth or a potential mate?

With relatively simple sets of goals (food, warmth, sex), and consequently only a limited amount of information to process from its environment, then fairly simple, autonomic, mechanisms can prove adequate for making choices. As soon as choices and possible responses become more complex, mechanisms for discriminating multiple inputs and goals are necessary and this seems to be the origin of rudimentary cognitive capacity — intelligence.

What evolves is what might be called a sort of 'behavioural jukebox',[2] a set of behavioural patterns — often contradictory — from which the jukebox operator can select in response to their environment and preferences. Moreover these behaviours aren't just selectable one at time, but can be played simultaneously (or near enough simultaneously). So our little male rove beetle can switch from playing 'hard to get' female towards a larger rival; to playing aggressive rival to another smaller

---

2] I got the idea of a behavioural jukebox from the excellent book by Bateson and Martin, *Design for a Life: How Behaviour Develops* (1999).

male or even courting male to a female; and then back to imitating a female almost immediately (fast enough that the larger male doesn't always notice). Perhaps it should be called a cove beetle? Of course, such strategies don't always work — rove beetles pretending to be the 'hard to get' female sometimes get discovered and attacked by their larger rivals. But that's life for you — nothing's certain.

The rove beetle's tiny 'behavioural juke box' provides it with a small amount of behavioural flexibility, which is nothing compared to the adaptable behaviours exhibited by the more intelligent mammalian species such as cetaceans, elephants and above all primates. The history of evolution of these species is, in part at least, the gradual accumulation of more and more behavioural 'modules' which give them greater adaptability, greater room for flexible responses to their environments (within their physiological limitations).

How might such systems evolve? In *The Sciences of the Artificial* the late Herbert Simon gave one possible answer — what he calls 'nearly decomposable' (ND) systems (Simon 1999). He illustrates this idea with a parable. Imagine two watchmakers assembling watches which are each made of 1,000 pieces. One watchmaker has to assemble each and every 1,000 parts, one after the other. The other has a design which is made up of ten sub-assemblies of 10 further sub-assemblies. Which watchmaker finds it easier — given the possibility of interruptions and errors — to assemble her watches? The answer is pretty obvious.

The idea of modules is not new — evolutionary psychology is almost entirely based around the idea of modular evolution of the mind (Barkow, Tooby et al. 1995). The idea is less often explored in terms of social behaviour patterns, although it provides an obvious explanation for the sort of variable inherited behavioural patterns obvious even amongst primate species (see for example de Waal 1989; 1996). The idea of conflicting, paradoxical, behavioural modules is (I think) pretty novel — at least expressed so explicitly. Yet it seems to offer a useful way of understanding otherwise perplexing behaviours.[3]

3. Yobs, Hippies and Paradoxical Primates

Group selection may have had an impact on how paradoxical social instincts became confirmed aspects of human nature — indeed given just how much of social species we are it would be surprising (assuming

---

[3]   I hasten to add this does not imply these are spacially located modules in the brain — even those which were thought to be so organised in the past, such as speech, are now proving to be rather more complex than originally thought. Nor does this imply a 'gene for aggression' or other such nonsense.

one accepts group selection can be part of the evolutionary process) if it were not so.

Let's consider this possibility through a little, only slightly tongue-in-cheek, 'thought experiment'. Let us imagine three bands of proto-humans evolving on the African steppe. We will call our first band the Yob Primates (YP) — they have very strong tendencies towards selfishness, aggression, competition and individualism. Our second band we will call the Hippy Primates (HP) — they are strong on altruism, peace-making, cooperation and conformity.[4] Finally we have our Paradoxical Primates (PP) — they are a confused bunch, capable of selfishness and altruism, aggression and peace-making, competition and cooperation and finally autonomy and conformity.

Remember we are considering here proto-humans, primates, not unlike ourselves and our closest relatives, the chimps, in many respects. They are omnivorous, physically similar to us (with all the advantages and disadvantages of our physique), reproduce in a similar way (young need prolonged rearing), etc. They live in an environment in which food is relatively easy to obtain, but is seasonal and sometimes well spread out and there are dangers from a number of predatory species and even from their own kind.

Amongst primates more generally it is not difficult to find examples which come close to both extremes, at least in regard to some of these instincts. Bonobo's have turned 'make love not war' into a life-style most Hippies can only dream about whilst some other primate species make Yobs seem positively peaceful.

So — how would our three groups get on? Well, firstly it is difficult to see how our troupe of Yobs (YP) could even be called a group, let alone a social group. Their pronounced instincts for aggression, competition, selfishness and individualism would lead to a very atomised troupe indeed. They would certainly have the edge at an individual level over, say, the Hippies, but remember we are talking here about group selection. As a group they would be pretty dysfunctional however advantageous their traits might be at an individual level. It is possible that a hierarchy capable of directing the group to act collectively might emerge, but reinforcement costs for the leaders would be immense. And social cohesion is important to a species which has the individual limitations of primates (physical limitations compared to predators; the need to both hunt and gather; and the need for long periods of child rearing, etc.).

---

[4]   The last may seem a bit odd, but remember how easy it is to recognise a Hippy on sight and you realise how conformist they are.

The Hippies (HP) on the other hand would function extremely well within their group. Their emphasis on cooperation, altruism, peace-making and conformism would lead to very strong internal cohesion. Their downside would be two-fold: firstly, they wouldn't be very good at achieving leadership within the social unit — with an aversion to competition and a liking for conformism, who decides when the troupe should hunt, or move to a new area to gather? This might be done collectively but this has a high communications cost. Secondly, their lack of aggression could prove a real handicap when it came to defending themselves against external threats — whether from predators or other groups.

Our Paradoxical Primates (PPs) could, just possibly, have the best of both worlds. In such a group some individualism, competition and aggression could lead to hierarchy emerging but conformist, cooperative and peace-making instincts would make the costs of maintaining cohesion relatively low. Peace-making and cooperation directed internally could be coupled with aggression and competition directed externally — rather like a successful football team (a very human invention).

One crucial characteristic of the PPs would be a tremendous adaptive advantage: the capacity to innovate. The individualist YPs could of course innovate — come up with new ideas about how to do things differently — but given their relative social dysfunction implementing any change which involved spreading either social innovation (e.g. division of labour) or even technical innovation (e.g. new tools or ways of hunting or finding food) would be difficult to implement. The HPs on the other hand would find it difficult to innovate at all. Our Paradoxical Primates could do both — invent new ways of doing things — socially or technically — and adapt to actually do them.

There is striking parallel here with the origins of life itself. As Robert Winston (2002) points out, DNA is a very paradoxical substance. A self-replicating molecule which was 100% effective and never made mistakes could never evolve. It would always stay exactly the same. So if DNA had evolved as a perfect replicator which never made any errors life on Earth would consist of a soup of DNA-like molecules and nothing else. On the other hand, if DNA was too inaccurate in replicating itself it would never survive at all — after only a few replications you would end up with garbage — a bit like making too many copies of copies with poor quality video-tapes. So DNA needed to have a 'just so' quality — not too perfect, but accurate enough to be nearly right every time. (It also needed some redundancy and self-correction protocols, but we won't go into that here). In other words — continuity and a little bit of change is what's needed to make evolution work (as well as selection of course).

So our little troupe of Paradoxical Primates could, if they had the right balance of paradoxical instincts and, perhaps just as importantly, could deploy the right ones at the right time (which requires intelligence), have great adaptive advantages over less flexible competitors. Such a delicate balancing act is difficult to achieve and there are certainly plenty of examples from human history of it going dreadfully wrong — both individually and collectively. It could even be argued that human madness might be, at least in part, a consequence of the stresses generated within individuals by paradoxical instincts. Interestingly, we even refer to what we consider some 'insane' behaviour as 'inappropriate' — that is the behaviour in itself is not 'wrong' but it is done in the wrong context. When paradoxical traits are used constructively, however, they can prove to be a huge advantage. Our Paradoxical Primate band would, potentially at least, certainly have the edge over both the Yobs and the Hippies.

This little thought experiment is not meant to be taken too seriously. It is supposed to be illustrative rather than offer an actual hypothesis about how humans evolved paradoxical social instincts. Actual human evolution from our primate ancestors was surely an extremely complex process and their were many pathways explored, as the confusing array of proto-human types suggests (Leakey and Lewin 1992). It is also pretty obvious that paradoxical social instincts pre-date humanity — too many of our primate cousins exhibit similar traits not to believe that our common ancestors had them too, at least in some degree.

We have also deliberately left out a crucial factor — the role of culture. We are now more aware than ever that culture is not a uniquely human trait (Bonner 1980) but humans are clearly much more adept at creating and transmitting culture than any other species. We have already noted how human culture can itself be paradoxical — especially in the practical, religious and moral proverbs we construct for ourselves. But it is obviously also the case that cultures can create, or reinforce, cultural differences in the emphases which are placed upon different paradoxical human traits.

The national cultural differences analysed in Hofstede's (2003) work already suggests that this is the case. Frans de Waal suggests that, for example, the cultures of countries with high population densities, such as the Netherlands, Japan and Java, may produce cultures which emphasise the tolerant, conformance and consensus whereas 'populations spread out over lands with empty horizons may be more individualistic, stressing privacy and freedom instead' (de Waal 1996 p. 201). Culture is therefore, in its more social norms and moral dimensions, a malleable result of underlying tensions in human social nature.

4. Paradoxical Instincts and Social Formations:
Fission-Fusion Societies

So far human social formations have been, by implication, considered as
relatively stable entities. Our little bands of Hippies, Yobs and Paradoxical Primates are, by assumption, stable bands. But humans do not live in
this way, and some of our closest primate relatives exhibit a rather different pattern — what are now being called 'fission-fusion societies'. These
are forms of social organisation in which groups follow patterns of
grouping together and splitting apart and coming together again in new,
or sometimes the same, groups. This is a very widespread phenomenon
in the animal world and usually it is easily explained by clear environmental factors to do with feeding, sex or external threat, or sometimes
combinations of these (e.g. fish and birds which migrate in shoals and
flocks to breeding grounds).

However, this is to concentrate on the 'fission-fusion' aspect of (some)
animal behaviour and ignore the 'societies' bit. It is where animals form
genuinely 'social' groups that this becomes most interesting and studies
of the more intelligent animals such as primates and cetaceans have been
the most intriguing. These species engage in complex social interactions
but, and this is the crucial point, their social groups are not necessarily
permanent (unlike say colony species such as ants and termites). This
gives their social groupings immense flexibility and adaptability, a clear
evolutionary advantage.

The sorts of issues which come up are why and how do social groups
form and fragment? How is social behaviour created, or rather recreated,
in such fluid circumstances? And how do individuals cope either with
being isolated or with (re)entry into different social groups?

The implications for human behaviour are also important because, as
Stanley Milgram (see section 4.3) has pointed out, humans have evolved
to be able to operate both as isolated individuals and, predominantly, as
members of social groups. Human social groupings appear to be an
extreme form of fission-fusion societies with their extreme flexibility and
relative ease of decomposition and (re)creation.

Fission-fusion has costs associated with it. Constantly having to create
or re-create social bonds requires greater investment than more permanent arrangements. Inauspicious fissions or fusions occurring at environmentally inopportune moments can be disastrous. Fissions may also
involve enforcement and inter-group conflict costs. So, a fission-fusion
strategy must have commensurably high benefits to justify its evolution
in a number of primate groups, including humans. So what is gained?

It is useful here, to illustrate the benefits of a fission-fusion social strategy, to engage in a little thought experiment of what human societies

would look like without this fission-fusion capability — i.e. what if they were permanently fissioned or fused? A permanently atomised human social group would be incapable of the feats of collective productivity and flexible response to their environment which are evident in even the most early proto-human societies. On the other hand, a permanently fused social grouping would suffer rigidities, introversion, possible in-breeding and other effects of lack of flexibility which a fission-fusion pattern can avoid.

Human social evolution seems to have started out as very much a fission-fusion structure. There is a parallel here with chimps, who are not only our closest relatives but, we now know from the pioneering work of Jane Goodall and others, very much live in fission-fusion societies in the wild. But when humans became 'domesticated' through the invention of agriculture (Wilson 1988) we began to live in more stable social groups, with much less mobility between groups. Similarly, in captivity chimps cannot engage in fission-fusion activity — like 'domesticated' humans they become more fixed in their social groups. Frans de Waal argues that both chimps and humans may be 'pushing our adaptive potential to the limit' but that we can both survive successfully in permanent groups — an example of our (chimps and humans) immense 'psychological and social plasticity, which in and of itself is a natural ability' (de Waal 1996 p. 169).

Modern human society seems to have reverted to the fission-fusion model, but with the added twist that we belong to numerous 'fusions' at the same time (at home in our communities, at work, professionally, socially, etc.). The idea of fission-fusion has so far been limited to ethological studies, but it may have great potential in studies of modern human social, institutional and organisational systems. Recent work on the idea of 'social capital' — i.e. the density of social networks of trust and reciprocity — suggests that in modern human societies this is an important factor in their success (Putnam 1993; Dasgupta and Serageldin 2000). This would seem to have strong links to the idea of fission-fusion social behaviour. Likewise, Herbert Simon's work (Simon 2000) on the idea of 'nearly decomposable' organisations has strong potential links to the idea of fission-fusion societies. This is clearly an important issue which social scientists — especially in applied areas like public policy and organisational studies — could probably benefit.

Chapter Six

Climbing Mount Paradoxical

The 'paradoxical primate' model developed in these pages may offer, just possibly, a way of reconciling some of the fissions within social sciences and between them and the natural sciences — a new fusion which E.O. Wilson (1998) has called 'consilience'. Most often this has been seen as a fusion in one direction — from the natural sciences to the social sciences. There is a great deal to be said for this — social sciences have, until very recently, been mostly separated from and ignorant of — not to mention hostile to — developments in fields such as biology, ethology, primatology, evolutionary psychology, behavioural ecology, etc. What I hope I have demonstrated in these pages is that there is some thinking, at least, within the fields of economics, public administration, organisation and management and related studies about the paradoxical nature of human behaviour which could be profitably integrated into evolutionary science. The traffic does not need to all be one way.

In this final Chapter I do not want to summarise all the arguments put forward. Instead I will ruminate on where we might take this new approach. The aim of the whole book and especially these concluding conjectures and hypotheses is to stimulate debate. What is being offered here is not a complete theory but rather a 'just so story' in the very best sense of the term 'hypothesis' — a testable, refutable, reasonable, guess at what might explain our nature a little better. Some of these thoughts will be deliberately self-critical — putting up some of the arguments which might be used to criticise my whole approach or particular aspects of it. As the Jesuits know, there's nothing like a bit of dialectical debate to sharpen ideas up.

(1) Firstly, I want to re-examine briefly the idea that we may have evolved (paradoxical) social instincts — and the alternative 'blank slate' theory.

(2) Secondly, I take a self-critical and, if I am honest, self-defensive look at the particular model of evolved paradoxical social instincts proposed. The model has its faults — probably more than outlined here — but I argue it also has its uses.

(3) Thirdly, I say something briefly about the dynamics of paradoxical human systems — how they seem (paradoxically) to be both patterned and almost random in their changes.

(4) Fourthly, I say something about the inter-play between instincts, institutions and intelligence which are the three fundamental forces shaping human behaviour. In some ways this is the core of the whole argument in this book.

(5) Fifthly, and finally, I want to take another brief look at the whole idea of 'consilience' and how the ideas developed in these pages might help facilitate one small aspect of such an endeavour — and what some of the obstacles are.

1. Evolving Paradox

The idea of any genetic and evolutionary component to human behaviour and our resulting institutions has been, to put it mildly, just a little controversial. There are those who resolutely deny the possibility that humans have any biological component at all to our behaviour, or at any rate none of any significance — we are 'socially constructed' not biologically:

> What we in fact construct is our cognitive, conscious selves, that part of ourselves that knows what's what and who's who. There is a clear-cut denial in both Mead and in Berger and Luckmann (1975) of the idea that we are biologically limited to certain forms of social organization. The latter write 'there is no human nature in the sense of a biologically fixed substratum determining the variability of socio-cultural formations'. With this I agree. *Arguments that there are biological determinants of 'social formations' are unacceptable.* Organic inputs go so far as indicated in earlier chapters and no further. It is the inputs from social history via its ongoing institutions and via the ideas and knowledge systems they perpetuate, reaching our neurones via significant others with whom we interact, that very largely make us the people we are, doing the things we do and making the social arrangements we make, not our bodies or the organic construction of our brains. (Reynolds 1976 p. 199; emphasis added)

Note the language here — 'arguments that there are biological determinants of "social formations" are *unacceptable*'. Not empirically or theoretically unsupported or unproven — they are 'unacceptable'. I think most people would see this as a declaration of a value judgement rather than a scientific statement.

Reynolds' was writing just a year after Edward O. Wilson's *Socio-biology* (1975) was published nearly 30 years ago. Much more recently some steadfast defenders of these ideas are fighting their corner with, it has to be said, increasing desperation (Rose, Lewontin et al. 1984; Lewontin 1991; Rose and Rose 2000). Whether these stalwarts like it or not, the idea that there is indeed a substantial genetic and evolutionarily based component to human behaviour has had mixed fortunes but it is now fast becoming an accepted, if not quite triumphant, view (Alcock 2003). Hence their desperation.

There continue to be intense debates within the general camp of social–biological explanations but it is undoubtedly much more accept-able. The popularity of Steven Pinker's recent best-seller *The Blank Slate* (2002) is a far cry from the time when Wilson was being doused with iced water at conferences and condemned as a Nazi. Fortunately we are now in an era where the most intense debates are about how we evolved to be how we are, rather than whether we did so evolve.

The simple thesis we have been developing in this book is that humans have innate instincts which are paradoxical in nature and have evolved as a consequence of our (fairly) unique form of social organisation with its combination of fission-fusion groupings and individuality. These paradoxical instincts combine with our truly unique intellectual gifts to produce amazingly adaptable beings and institutions. But as individuals and collectives we remain bounded by the contradictory pulls of con-flicting instincts, which may be for a time overcome but always remain as shaping forces in individual and collective behaviour.

We have evolved social mechanisms for coping with our paradoxical natures — including an elaborate capacity for individual, organisational and social hypocrisy — but in our recent history we have failed to explore scientifically this feature of our existence due to an overwhelm-ingly simplistic rationalist dogma.

Paradoxical human nature is an evolved trait that is highly adaptive for social animals, providing for a broad spectrum of possible behav-iours within a paradoxical system of possible behaviours. Individual humans may inherit greater or lesser propensities (but not deterministic impulses) towards one or other 'pole' of a paradox but few (if any) indi-viduals will be exclusively biased in one direction.[1] Cultures may encourage expression of one part of a paradoxical pair (e.g. aggression or competition) whilst discouraging expression of others (e.g. peace-making or cooperation) but they are incapable of permanently suppress-ing the opposite pole. Hence the dynamism and diversity of cultures, but

[1]   There may be cases — e.g. an aggressive sociopath — where there is a strong bias. But most of us are far too conflicted to be that simple.

within a set of species-specific boundaries defined by impossibilities (e.g. perfectly, permanently, competitive or cooperative societies are not possible, but strongly competitive or cooperative ones are, but they are also liable to be unstable in the long term).

2. Modelling Paradoxical Instincts Redux

The model of paradoxical human instincts outlined in Chapter 4 (repeated below as Figure 9) was deliberately formulated around a model of stylised contrasting paradoxical elements (see Figure 6/9) which has become fairly common in analyses of such phenomena (e.g. Quinn 1988; Quinn and Cameron 1988). This model has the advantage of simplicity but it is a theoretical construct rather than derived from empirical analysis. It may well be too rigid, or simple, in its current form but it does highlight the genuinely contradictory and paradoxical nature of these evolved instincts. Let me anticipate some possible criticisms of this representation by admitting that there are indeed several obvious drawbacks to the use of this rather stylised model.

Firstly it suggests a balance, or equality along each individual pole. If we take the case of autonomy versus conformity as an example, then Stanley Milgram's empirical evidence would tend to suggest that most people (in his limited US sample) tended to be stronger on conformity

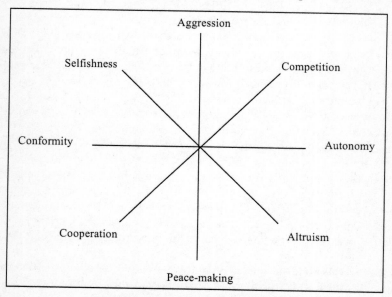

Figure 9: Paradoxical Human Instincts?

than on the individuality dimension. More people agreed to carry on shocking their 'victims' than rebelled, although the results also showed that even conformity had its limits and many would eventually rebel if pushed far enough. Geert Hofstede's work on national cultural differences (Hofstede 2003) included a dimension called 'individualism' which is defined as:

> Individualism (IDV) focuses on the degree the society reinforces individual or collective achievement and interpersonal relationships. A High Individualism ranking indicates that individuality and individual rights are paramount within the society. Individuals in these societies may tend to form a larger number of looser relationships. A Low Individualism ranking typifies societies of a more collectivist nature with close ties between individuals. These cultures reinforce extended families and collectives where everyone takes responsibility for fellow members of their group.[2]

On this individualism index (which is of course not quite the same as our 'individualism–conformism' dimension but is close enough to be useful) the USA — site of Milgram's work — scores a 91. Other 'Anglo' countries' have similarly 'high' scores — indicating a more individualist culture — UK 89, Australia 90, New Zealand 79, Canada 80. Asian countries have much lower scores — China 20, Hong Kong 25, Taiwan 17, South Korea 18, Japan 46, India 48.[3] (Interestingly the Sinic countries — China, Hong Kong, Taiwan — all have similar scores despite their very different political-economic systems).

It would be a reasonable hypothesis that if Milgram's experiments were repeated in countries with a less individualistic culture than the USA (where there was more conformance than rebellion) the results would be even more skewed towards 'obedience to authority'.

So 'autonomy' and 'conformism' are probably not evenly balanced and may be influenced by a wide range of factors — individual inheritance, institutional and cultural context, personal experience, etc. This does not however obviate the idea that in all cultures both individualism and conformism co-exist. Indeed Hofstede's results demonstrate precisely this — both individualism and conformity exist in all the countries he has surveyed — the only differences are in the balance between the two. I would also add that because his methodology is of a fairly standard 'mutually exclusive' type it probably misses the possibility of actual paradox — for example countries which might score highly on both 'individualism' and 'conformism', a really intriguing thought. Japan's individualistic attitude to religion (mentioned above), for example, contrasts with its otherwise highly conformist culture and might

---

[2] http://www.geert-hofstede.com/geert_hofstede_resources.shtml
[3] http://www.geert-hofstede.com/hofstede_dimensions.php

explain its rather 'in-the-middle' score on Hofstede's individualism dimension.

Secondly, by emphasising only contradiction the model tends to belittle the positive relationship or complementarity which can exist between opposing pairs. In Bob Quinn's work on paradox in organisations (Quinn 1988; Quinn, Faerman et al. 1996) he frequently points out that what is important about these permanent contradictions is how they are managed. That they are contradictory is obvious, but as Quinn and his colleagues and others (such as Collins and Porras 1994) have pointed out such contradictions can be put to positive use — they can generate 'creative tension' which help propel people and organisations to high levels of performance. On the other hand, if not handled correctly, they can also lead to destructive conflict.

Thirdly it suggests a degree of equality between the four pairs of paradoxes. In a recent discussion with Japanese colleagues about the model of human motivation proposed here one colleague (Prof. Tsukamoto of Waseda University) suggested that the dimension encompassing what he called 'part of the whole' versus 'autonomous individuals' might not be more fundamental than the other dimensions. This seems to me a very fair question but it is one which could only be answered by a great deal more theoretical and, crucially, empirical work in this area. My hunch would be that it is indeed probably the most fundamental dimension — it is what distinguishes us humans from either fully independent species on the one hand and fully integrated, colonial, social species on the other. This clearly relates to our earlier discussion about 'fission-fusion' societies.

Fourthly, the model emphasises the contradictions between pairs and by implication downplays the differences and contradictions between other elements. It could certainly be argued that one set of poles — aggression, autonomy, selfishness and competition — would in most people's thinking naturally group together and represent the opposite of the remaining four aspects — peace-making, conformity, altruism and cooperation. Indeed many would see these as representing the fundamentally opposed visions of human nature espoused by the traditional 'right' and 'left' in politics. The reality is of course far more complex. Aggression, for example, can be used to defend or suppress autonomy, for cooperative or competitive and for selfish or altruistic ends. Likewise peace-making might serve the purpose of ensuring conformity or allowing individuals to act autonomously. And so on. Each of these elements can be contradictory to every other element, even if they tend to be more often juxtaposed as in this model.

Fifthly the model, of course, also restricts our analysis of paradoxes to just eight items — whilst this may be aesthetically pleasing and conform to some rather ancient ideas about how to analyse such things (in Taoism for example) it is still restrictive. Moreover it stresses bipolar paradoxes whereas, as the previous point suggests, contradictory relationship may be far more complex than that. Other much more fluid methods can be used to 'map' paradoxical or contradictory elements within systems (see for example Rosenhead 1989; Huff 1990). The utility of using a more formal, rigid, model is less about exact maps of actual systems and more about getting across how they are structured — the approach is more conceptually communicative and less representational. All models are after all just models and they can serve different purposes. For communicating the basic ideas of paradoxical human social instincts this model seems useful.

3. Dynamics of Paradoxical Systems

Paradoxical human systems are themselves paradoxical — they exhibit simultaneously tendencies towards patterned behaviour and to be chaotic; they are both bounded and highly adaptable; they are simple but also very complex; they can appear both stable and highly dynamic; and so on.

Bounded Adaptability

Paradoxical human social instincts and paradoxical human institutions are both bounded and highly adaptive. They are bounded because — whilst individual humans may, very occasionally, step widely outside of the range of 'normal' behaviours — as a totality human behaviour, including within our institutions and organisations, is bounded by the limits imposed by permanently conflicting impulses.

This has enormous implications for how we run our organisations — which have been extensively explored, if not widely enough accepted (see Chapter 1). It has similarly large implications for how we run our societies and governments and this is, as yet, nowhere nearly as well explored (see Chapter 2). But it is possible to map out some tentative conclusions from our analysis. There always seem to have been those within human societies who felt it was possible to 'perfect' these societies, or at any rate some individuals within them. For most of human history this was expressed in religious terms but from about the seventeenth century onwards it began to be articulated in politico-economic language. One conclusion from our analysis is that certain types of human perfectibility

— e.g. the perfectly cooperative or perfectly competitive society — is almost certainly impossible to achieve.

This seems supported by twentieth-century experience — not just the massive experiment with so-called socialism in the USSR, China and other countries, but also smaller scale attempts such as the kibbutz movement in Israel or western Hippy communes of various types, have all collapsed eventually under the weight of their internal conflicts. If the collapse of the Soviet Union (and the despicable crimes carried out in the name of socialism in the USSR, Cambodia and elsewhere) tell us anything it should be that utopian attempts to create the 'New Man' are doomed to failure.

Does this mean the 'end of history' and that liberal democratic capitalism is the only viable model of social organisation? The answer to this is probably both 'yes' and 'no'. It certainly seems at the moment that the many and varied forms of democratic capitalist regimes are proliferating around the globe and that dictatorships — whether supposedly utopian or just plain kleptocratic — are in retreat. But there are several caveats which suggest that a simple 'convergence' thesis — we are all going to end up with some sort of liberal or social democratic state somewhere on the spectrum between Sweden and the USA — is highly unlikely. The adaptability of the paradoxical human social instincts is both a massive strength and a weakness. The adaptability and inherent instability it produces makes adaptation possible and conversely it makes stable social forms highly unlikely or at any rate problematic.

Human systems are almost certainly always going to be in a state of flux and evolution, precisely because our human natures make us adaptable. We will constantly be evolving different politico-social-economic patterns but within (mostly) certain limits. These limits are that human nature will not permit certain types of pure regime — such as a perfectly integrated 'hive' like human system. (That only happens in science-fiction — e.g. the Borg in Star Trek. Such fictional examples, if one cares to think about them seriously as sociological thought experiments, actually help to illuminate just what the natural limits of human social organisation really are.)

It is even probable that attempts will arise again to create some form of utopian society. Indeed even as you read this, one movement — fundamentalist Islam — has just such an objective (although their utopia is rather dystopian from most perspectives, including a majority of Muslims). The rise of some form of collectivism again cannot be ruled out. Nor can the rise of attempts to create more anarchic, perfectly competitive societies — it is likely that a good portion of humanity already lives outside of effective state control for a significant part of their existence

(Schneider and Enste 2003). It seems to be a characteristic of humans to want to achieve some sort of stable, consistent, utopia and to constantly fail to realise it.

Patterned and Chaotic

Numerous philosophers and social scientists over the centuries have tried to come up with cyclic theories of human history — most often with regard to the rise and fall of empires. More recently management writers have come up with similar theories about business organisations (e.g. Miller 1990). These theories have always proved attractive and seem to offer tantalising glimpses of underlying patterns in the tides of human affairs. And yet, their promise never quite seems to be delivered. A bit like weather forecasters who always seem nearly right until the unpredicted hurricane comes along, such social theories seem 'always the bridesmaid but never the bride'.

Simplicity and Complexity

The most promising approach to solving this problem — or understanding it at least — seems to have come from chaos and complexity theory. This is a bundle of ideas which even its most enthusiastic proponents admit is still in its early days and is not yet a theoretically coherent or integrated body of knowledge. It already seems to offer a way of modelling and understanding the dynamics of at least some complex systems, including human ones (for the best introductions see Holland 1995; 2000; Kauffman 1995; Axelrod and Cohen 1999). What chaos and complexity theory identifies are systems which can produce forms of 'order out of chaos' which seems both random and patterned at the same time — similar but not quite identical formations; unpredictable cycles; simple ingredients forming extremely complex wholes; all of these characterise these systems and seem highly relevant to paradoxical systems as well. Even relatively simple paradoxes can — through a process of combination — lead to extremely complex adaptive systems. There is an obvious parallel here with the Taoist *pa-kua* arrangement of yin and yang used in the *I Ching* to create a large number of combinations from a small number of possible states. Indeed the Taoist philosophical tradition in general and the Yin-Yang school in particular say a lot of things which seem to parallel the paradoxical human social instincts approach.

It also seems that paradoxical systems may tend towards settling into 'quantum states' analogous to 'strange attractors' in complexity theory. That is, although the range of possible combinations in a paradoxical system is infinite, in practice only a sub-set of combinations and configurations tend to be identifiable for substantive periods. These quantum states are probably system specific and can only be established empirically by (inductive) observation of actual systems dynamics. While there are some explanatory models emerging which might have predictive possibilities (e.g. the 'interacting agents' model in economics [Ormerod 1999]) these systems are so inherently complex that only empirical research is likely to be of substantial use for the foreseeable future.

Stable and Unstable

However stable the system appears, for however long, there is always a potential for small or large shifts in the configuration of the system. Sometimes only relatively minor events can trigger major shifts — as in 'catastrophe theory' where small incremental changes eventually trigger 'catastrophic' systems changes. What makes these systems 'paradoxical' is the simultaneous existence of inherently incompatible elements and while they may sometimes appear more 'rational' and 'stable' — for a time — the potential for change is always present even if not actually happening at this moment.

The so-called 'butterfly effect' is now well known — the idea that a butterfly flapping its wings on one side of the planet can so affect the global weather system that it changes the course of a hurricane on the other. This 'just so' story (no-one, as far as I know, has ever measured such an event) is meant to illustrate that certain types of complex systems are highly susceptible to 'initial conditions' — only a tiny change in their starting points can cause large-scale divergences in their trajectory, many orders of magnitude larger than the initial differences. What might be called the 'reverse butterfly effect' is less often discussed — that is that complex adaptive systems can sometimes absorb large amounts of additional input without altering substantially. Whether such systems react to small changes or resist large ones is a very tricky area and is what, in part, contributes to their inherent specific unpredictability. Paradoxical human systems seem to have the same characteristics — they can, for example, appear highly resistant to change for very long periods only to suddenly change dramatically from seemingly quite small changes. The collapse of Soviet Communism which survived the massive wars of intervention in the 1920s, huge social reorganisation and industrialisation in the 1930s, Nazi invasion in the 1940s, armed rebellions in the 1950s and 60s and the Cold War, only to apparently surrender to a few thousand peaceful demonstrators and a bit of additional weapons competition, seems a good example.

Adaptive and Maladaptive

We have stressed the adaptability of paradoxical human social instincts and tended to assume that such flexibility is always positive. Unfortunately the hypothesis is that paradoxical systems are potentially both highly adaptive and maladaptive. The contradictory nature of these instincts, and the institutions they give rise to, means that it is just as easy to select the wrong response as the best one.

Moreover learning, for individuals, institutions, cultures and even across generations, is not necessarily a good guide to action. As Danny Miller has pointed out in his book *The Icarus Paradox* (1990), in relation to human business organisations, there is all to often a tendency to rely on what has worked in the past, when changed environmental circumstances may mean these lessons have become obsolete. As with biological mutations, the distinction between 'adaptive' and 'maladaptive' change cannot always be predicted in advance. And, like biological mutations, what worked *then* might not work *now*.

This should not be taken to imply that there cannot be social learning. Humans have experimented with a bewildering (to us) range of social and cultural arrangements (although this range is less great than we often suppose looking at them from the 'inside'). During these experiments we have gradually accumulated knowledge about 'what works'. Human life in the advanced capitalist democracies is incomparably more peaceful, safe, healthy, and materially, educationally and culturally affluent for the majority than ever before in human history. This is not to be complacent about any of the evident problems within these societies or globally. But what is obvious is that we now know at least how to get some things 'right' about how to organise our very social species in ways which benefit the majority if not all of us.

Unfortunately this triumph has been more of a practical one than a theoretical advance — social sciences have made nothing like the strides accomplished by the natural sciences. The old dictum 'it works in practice, now let's see if we can make it work in theory' never seemed more apposite.

Paradigms and Paradoxes

I just want to take one final pot-shot at the idea of 'paradigm shifts' in human social, political and business systems. A paradigm shift is, I would suggest, a shift in the underlying structure of paradoxes. For example a shift from the 'centralise and decentralise' paradox to a 'hierarchy and network' paradox would represent a genuine paradigm shift. All too often what pass for 'paradigm shifts' in much social scientific discourse are nothing of the sort — they are merely a shift in the balance between one or more underlying paradoxical structures.

A paradoxical system reconfiguration is often mistaken for a 'paradigm' shift. A shift from 'centralisation' to 'decentralisation' is a deceptive (but real) change within a 'paradox space' which may suggest paradigmatic change where none has actually occurred. A lot of organisational change in the in the 1980s which was supposed to be a paradigm shift to supposedly 'post-bureaucratic' organisations was, as one pair of writers memorably put it, a highly deceptive change. What was actually being created, they suggested, was more often than not a 'cleaned bureaucracy' — i.e. one which had successfully reverted to the original foundational principles of bureaucracy and managed to shed the accumulated detritus of decades of ossification (Heckscher and Donnellon 1994).

4. Paradoxical Instincts, Institutions and Intelligences

How do paradoxical human instincts interact with both our institutional/cultural environment and our obvious intelligence? I want to suggest that there is a dialectical interaction between these three primary elements of humans: our individual, paradoxical, instinctive nature; our paradoxically informed institutional and cultural environment (which shapes and is shaped by us); and finally by the (bounded) rational choices we make both over competing instincts and values and over the less complex dilemmas and choices we confront. I have attempted (Figure 10) to suggest how these 'map' onto human behaviour.

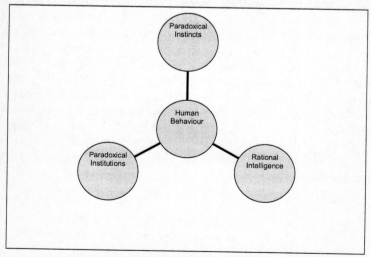

Figure 10: Instincts, Institutions and Intelligence

I should stress that I am not suggesting that these three factors are of equal importance. It has been generally suggested in the behavioural genetics literature that there is a 50/50 split between inherited and learned components of behaviour (Harris 1998; Wright 1998). If this were true, then the paradoxical instincts component would clearly carry more weight than the remaining two. However, it seems to me that the whole '50/50' notion is too simplistic when dealing with such highly interactive systems as humans possess inside their skulls. What would be really needed to establish the precise way the contending forces mapped in this diagrammatic way interact would be a research programme dedicated to exploring empirically and theoretically just how they work in practice and could work in theory, which is beyond the scope of this little book. In the final section I try to map out very briefly what such a research programme might consist of.

5. Towards Consilience: The Role of a Paradox Theory?

In this final section I want to map out what the research agenda ought to be if the ideas synthesised in this volume are to be taken seriously. Before spelling that out, I want to take a slight detour through a rehash of some of the arguments about scientific method, integration and consistency mentioned at various points in the book up to now.

In his book *Consilience*, E.O. Wilson argues for a reintegration of natural sciences with social sciences and for greater integration within social sciences. This is part of a debate about the nature of science and indeed about the nature of reality itself. On the one hand stand those searching for answers which try to unify various branches of knowledge on the assumption that the Universe is a unified place where one thing leads to another (though not necessarily in any simplistic, deterministic sense) (Buchanan 2000). At the other extreme stand those who believe in a Universe where various branches of knowledge are irretrievably separate and un-connectable — the 'dappled world' as one writer memorably calls it (Cartwright 1999). And at the extreme end of this un-connected and un-connectable Universe stand some of the deconstructionists and social constructionists who believe that even physical realities are purely subjective — usually by drawing on some half-baked understanding of quantum theory.

The current argument against 'consilience' usually runs something like this: systems are made up of components but systems produce emergent properties which are not discernable from analysis of their constituent parts. Complexity and chaos theories have finally 'proved' that reductionism — breaking a system down into its components — and

determinism — assuming the interaction of components can be predicted provided their nature and initial conditions are known — are a
false approach.

There is a grain of truth in this argument but there is also a wonderful
magician's trick. It is true that chaos and complexity theories have established that certain systems (but not all) are subject to extreme sensitivity
to initial conditions and in any case most real-world systems are to some
degree open systems and even minor external influences can bring huge
changes (the so-called 'butterfly effect'). What is utterly false is that we
can therefore say nothing about the behaviour of these systems through
an understanding of their component parts and the rules governing their
behaviour.

As the 'butterfly effect' in weather systems is the usual example given
let us just consider it for a moment. Does an understanding of the compo-
sition of an atmosphere (the types of gases that make it up, the average
amount of water vapour in it, etc.) and the influences upon it (sunlight,
gravity, rotation of the planet, etc.) enable us to predict the weather? No,
of course not. But it does create what might be called a 'possibility space'
of types of weather we might see.

At the ludicrous extreme we know that the atmosphere is unlikely to
boil off into space unless something fundamental changes (so that's a bit
of a relief then). We also know that 1,000 mile-an-hour winds are also
extremely unlikely (phew). We know that seasonal changes will take
place (due to the earth's tilt), that latitudinal variations in weather patterns persist over time whereas storms don't (unlike for example the
famous storm on Jupiter which seems semi-permanent). And these are
not just observations, we have a pretty good understanding of why there
are limits to what is likely in our atmosphere, even if we cannot predict
the exact weather in Milton Keynes on the last Friday of next month.

Our knowledge has indeed developed sufficiently that we can, however, predict roughly the weather over several days in advance and even
more crudely over several weeks. These predictions are probabilistic but
they are good enough for certain purposes (e.g. for farmers or health service planning). So complexity reduces the scope for accurate prediction,
it does not say we cannot through reductionist and probabilistic science
make some broad guesses about what will happen next in complex systems. It allows us to understand the possibility spaces of what might
happen, within certain parameters. It often allows us to discern what
likely patterns might emerge, even when we cannot say with certainty
which precise ones will at any moment. Importantly it often enables us to
predict that certain states of the system (so-called 'attractors' in complexity theory), however apparently stable, will eventually change.

Consilience then is not about some Newtonian, mechanical, hierarchy of precise reduction and determination. It is rather about understanding how one level of systems determines parameters for how the next level may behave — what possibilities there may be for emergence even when we cannot predict their exact nature. So physics determines the possibility space for chemistry, which in turn determines the possibility space for biology, and so on.

Moreover we know that such systems are what in political science is called 'path-dependent'. That is, once a system has taken off in one direction — e.g. carbon based life forms — it cannot suddenly switch track and start utilising silicon as its basis. What we see is a sort of reverse cone of possibilities which in some respects gradually narrow as systems evolve, as they accrete more and more history and path-dependency. So whilst it is theoretically possible that out there somewhere there are silicon based life forms (we know it must be true because Star Trek told us so) we know that on this planet all life forms are (for the moment) carbon based. That in turn places certain limits of their possibility space (e.g. they don't survive very well in fire or indeed any heat above certain limits, etc.).

Here we come to an interesting paradox: within a limited possibility space there may be an infinite number of possibilities. That is, there may be an infinite number of small variations within a set of limiting factors. Carbon-based life has already produced a fantastic number of variations on a theme and these seem more or less infinite. So we have the strange paradox of limitation imposed by a path-dependent hierarchy of possibilities which is combined with infinite variety within the resulting possibility spaces. This is important for our arguments about paradoxical human nature because — I suggest — a fairly limited set of paradoxical human instincts can produce very large (if not infinite) possibilities for diverse behaviour and adaptability without the need for a 'blank slate' explanation for such diversity.

So I would suggest there seem to be three types of law-like relationships in reality:

- The first are what are being called **'ubiquitous'** laws by some — ideas from fields such as chaos, complexity, and network theories — which seem to work at a number of different levels of reality from inanimate matter to human systems in very similar ways (e.g. Holland 1995; Buchanan 2000; Johnson 2002; Gribbin 2004).

- The second level is what might be called **'consilient'** laws — that is relationships between different levels which clearly derive from laws at a lower level. This approach has worked extremely well in the physical sciences and there is a great deal of causal integration between physics, chemistry and biology (Wilson 1998).

- The third level are what have become known as **'emergent'** laws — that is, laws which operate at one level of reality and, whilst constrained from below, are not obviously deducible either from lower level rules or from 'ubiquitous' rules.

Human paradoxical instincts, it seems to me, probably have elements of all three 'law-like' properties. I have not explored the links to paradoxes in physical and biological systems, but I suspect they are there. I have no doubts at all that these paradox rules at the human level can be reconciled with lower levels. But I also believe that they also have emergent properties which are not entirely predictable 'from below' and therefore require specific research strategies to uncover them.

So what are the research implications of humans having (hypothesised) paradoxical social instincts? Let me suggest this in disciplinary terms.

Evolutionary psychologists have rightly posited an inherited human nature and have been dedicated to establishing what it is, how it evolved and how human brains have been evolved to embed it. In all three areas the 'paradox hypothesis' has profound implications. Much of the research on human nature, for example, up until now has tended to be within an 'either/or' frame, rather than the 'both/and' perspective implied by a paradox approach. We have already suggested the start of a guess at how group selection could have shaped the paradoxical instincts of early humans. We have no idea about how the mind might embed these paradoxical instincts and leave that to those far more knowledgeable to explore.

Behavioural geneticists have concentrated, when looking at humans, on establishing individual variability in inherited behavioural traits and their roots in the human genome. They already have results (on issues such as aggression) which suggest the causes of some variability. I am fairly confident that it will be possible for them, if they choose, to explore the variability on the elements of the model of human paradoxical social instincts suggested here.

Sociologists, political scientists and economists could all explore the relationship between instinctual traits, institutional influences and bounded rationality in the behavioural choices which we humans make.

Some economists and political scientists, in exploring 'beyond self-interest' (Mansbridge 1990), have already started on such a course (see also Chapter 2). Reconciling self-interested and altruistic behaviour has been deeply problematic in economics and the paradox approach suggested here offers one way, at least, of starting making sense of this conundrum.

Sociologists are especially well-equipped to extend the work of some of their colleagues on 'ideological dilemmas' (Billig, Condor et al. 1988) into a more sustained examination of the interaction between contradictory cultural and institutional norms and individuals. Some sociological theories about 'agency' and 'structure' would, I believe, take on a very different slant from a paradox perspective.

Organisational researchers — in both generic organisation and management and in the public administration field — already have a good start in examining paradoxical behaviours but this clearly needs integrating and extending.

For such a research agenda to succeed it would require several things. Firstly, widespread acceptance of the resolution of the sterile nature-nurture controversy through the broad movement that is sociobiology, evolutionary psychology, human behavioural genetics and ecology, etc. (Laland and Brown 2002). Secondly, it requires a shift in perspective by researchers in a range of different fields to accept the logic of paradox within their own theoretical and methodological frameworks. This does not mean, as I hope I have demonstrated, abandoning reason and science for some sort of mystical approach. On the contrary, the paradox hypothesis suggested in this book is thoroughly rooted in a realist and reasoned approach to understanding who we are and how we came to be this way. If it helps to stimulate this debate about how we might reconcile different levels of explanation of human behaviour from a paradox perspective I will be more than happy.

Bibliography

Alcock, J. (2003). *The Triumph of Sociobiology*. Oxford, Oxford University Press.

Ansoff, H.I. (1968). *Corporate Strategy*. London, Penguin.

Ansoff, H.I. (1991). 'Critique of Henry Mintzberg's "The Design School: Reconsidering the Basic Premises of Strategic Management".' *Strategic Managament Journal* **12**.

Ardrey, R. (1967). *The Territorial Imperative*. London and Glasgow, Collins Cleartype Press.

Arendt, H. (1963). *Eichmann in Jerusalem: A Report on the Banality of Evil*. New York, Viking.

Aucoin, P. (1990). 'Administrative Reform in Public Management: Paradigms, Principles, Paradoxes and Pendulums.' *Governance* **3**: 115–137.

Axelrod, R. (1990). *The Evolution of Co-operation*. London, Penguin.

Axelrod, R. (1997). *The Complexity of Cooperation: Agent-Based Models of Competition and Collaboration*. Princeton, Princeton University Press.

Axelrod, R. and M.D. Cohen (1999). *Harnessing Complexity: Organizatinal Implications of a Scientific Frontier*. New York, The Free Press.

Bailey, F.G. (1991). *The Prevalence of Deceit*. Ithaca, Cornell University Press.

Barkow, J.H., J. Tooby, et al., Eds. (1995). *The Adapted Mind: Evolutionary Psychology and the Generation of Culture*. Oxford, Oxford University Press.

Bateson, P. and P. Martin (1999). *Design for a Life: How Behaviour Develops*. London, Jonathan Cape.

Berger, P. and T. Luckmann (1975). *The Social Construction of Reality*. Harmondsworth, Penguin Books.

Berry, C.J. (1986). *Human Nature: Issues in Political Theory*. Houndmills, Macmillan Education Ltd.

Billig, M., S. Condor, et al., Eds. (1988). *Ideological Dilemmas: A Social Pychology of Everyday Thinking*. London, Sage Publications.

Bolman, L.G. and T.E. Deal (1991). *Reframing Organizations: Artistry, Choice and Leadership*. New York, Jossey Bass.

Bonner, J.T. (1980). *The Evolution of Culture in Animals*. Princeton, Princeton University Press.

Boston, J., J. Martin, et al. (1996). *Public Management: The New Zealand Model*. Auckland, Oxford University Press.

Brown, D.E. (1991). *Human Universals*. New York, McGraw-Hill.

Brunsson, N. (1985). *The Irrational Organisation*. Chichester, Wiley.

Brunsson, N. (1989). *The Organization of Hypocrisy: Talk, Decisions and Actions in Organisations*. New York, John Wiley and Sons.

Brunsson, N. and J.P. Olsen (1993). *The Reforming Organization*. London, Routledge.

Buchanan, M. (2000). *Ubiquity: The Science of History or Why the World is Simpler Than We Think*. London, Weidenfeld and Nicolson.

Cameron, K.S. and R.E. Quinn (1999). *Diagnosing and Changing Organizational Culture*. New York, Addison-Wesley.

Cannon, T. (1997). *Welcome to the Revolution: Managing Paradox in the Information Age*. london, Pitman.

Carley, M. (1980). *Rational Techniques in Policy Analysis*. London, Heinemann/PSI.

Cartwright, N. (1999). *The Dappled World: A Study of the Boundaries of Science*. Cambridge, Cambridge University Press.

Collins, J.C. and J.I. Porras (1994). *Built to Last: Successful Habits of Visionary Companies*. London, Harper Business.

Common, R., N. Flynn, et al. (1992). *Managing Public Services: Competition and Decentralisation*. London, Butterworth-Heinemann.

Dasgupta, P. and I. Serageldin, Eds. (2000). *Social Capital: A Multifaceted Perspective*. Washington, DC, The World Bank.

de Waal, F. (1989). *Peacemaking Amongst Primates*. London, Penguin.

de Waal, F. (1996). *Good Natured: The Origins of Right and Wrong in Humans and Other Animals*. Cambridge, MA, Harvard University Press.

de Waal, F. (2001). *The Ape and the Sushi Master: Cultural Reflections By a Primatologist*. London, Penguin Books Ltd.

Doray, B. (1988). *From Taylorism to Fordism: A Rational Madness*. London, Free Association Books.

Dugatkin, L. (2000). *Cheating Monkeys and Citizen Bees: The Nature of Cooperation in Animals and Humans*. Cambridge, MA, Harvard University Press.

Dunleavy, P. (1991). *Democracy, Bureaucracy and Public Choice: Economic Explanations in Political Science*. Brighton, Harvester/Wheatsheaf.

Elcock, H. (1991). *Change and Decay: Public Administration in the 1990s*. London, Longman.

Etzioni, A. (1988). *The Moral Dimension: Toward a New Economics*. London, Collier Macmillan Publishers.

Etzioni, A. (1993). *The Spirit of Community: The Reinvention of American Society*. London, Touchstone Books.

Etzioni, A. (1996). *The New Golden Rule*. London, Profile Books.

Fonseca, J. (2002). *Complexity and Innovation in Organizations*. London and New York, Routledge.

Foster, R. and S. Kaplan (2001). *Creative Destruction*. New York, Currency.

Geras, N. (1983). *Marx and Human Nature: Refutation of a Legend*. London, Verso.

Giannetti, E. (1997). *Lies We Live By: The Art of Self-Deception*. London, Bloomsbury.

Grant, R.W. (1997). *Hypocrisy and Integrity*. London and Chicago, The University of Chicago Press.

Gribbin, J. (2004). *Deep Simplicity: Choas, Complexity and the Emergence of Life*. London, Penguin Books Ltd.

Guba, E. G. E. (1990). *The Paradigm Dialog*. London, Sage.

Handy, C. (1995). *The Age of Paradox*. Cambridge, MA, Harvard Business School Press.

Harris, J. R. (1998). *The Nurture Assumption: Why Children Turn Out The Way They Do*. London, Bloomsbury.

Harvey, J.B. (1988). *The Abilene Paradox and other Meditations on Management*. San Francisco, Jossey-Bass Publishers.

Harvey, J.B. (1996). *The Abilene Paradox*. San Francisco, Jossey-Bass.

Heckscher, C. and A. Donnellon, Eds. (1994). *The Post-Bureaucratic Organization*. London, Sage.

Held, D. (1996). *Models of Democracy (2/e)*. Cambridge, Polity.

Hofstede, G. (2003). *Culture's Consequence's: Comparing Values, Behaviours, Institutions and Organizations Across Nations (2/e)*, Sage Publications Inc (USA).

Holland, J.H. (1995). *Hidden Order: How Adaptation Builds Complexity*. Reading, MA, Addison-Wesley.

Holland, J.H. (2000). *Emergence From Chaos to Order*. Oxford, Oxford University Press.

Hood, C. (1991). 'A Public Management for all Seasons?' *Public Administration* **69**(1).

Hood, C. and M. Jackson (1991). *Administrative Argument*. Dartmouth, Aldershot.

Hood, C., O. James, et al. (1999). *Regulation Inside Government*. Oxford, Oxford University Press.

Hori, I., F. Ikado, et al., Eds. (1972). *Japanese Religion: A Survey by the Agency for Cultural Affairs*. Tokyo, Kodansha International.

Huczynski, A. A. (1993). *The Management Gurus*. London, Routledge.

Huff, A.S.E. (1990). *Mapping Strategic Thought*. New York, Wiley.

Iacocca, L. and W. Novak (1987). *Iacocca: An Autobiography*. Toronto, Bantam Books.

Jardine, L. (1974). *Francis Bacon: Discovery and the Art of Discourse*. Cambridge, Cambridge University Press.

Jenkins, K., K. Caines, et al. (1988). *Improving Management in Government: The Next Steps*. London, HMSO.

Johnson, S. (2002). *Emergence: The Connected Lives of Ants, Brains, Cities and Software*. London, Penguin Books Ltd.

Katz, D. and R. Kahn (1978). *The Social Organizations of Organizations (2/e)*. New York, John Wiley and Sons.

Kauffman, S. (1995). *At Home In The Universe: The Search for Laws of Complexity*. London, Viking.

Kaufman, H. (1976). *Are Government Organisations Immortal?* New York, Brookings Institute.

Kelly, G.A. (1963). *Theory of Personality: The Pyschology of Personal Constructs*. New York, W.W. Norton & Co.

Kolakowski, L. (1999). *Freedom, Fame, Lying and Betrayal*. London, Penguin.

Kuhn, T. (1962). *The Structure of Scientific Revolutions*. Chicago, University of Chicago Press.

Laland, K. and G. Brown (2002). *Sense and Nonsense – evolutionary perspectives on human behaviour*. Oxford, Oxford University Press.

Lawrence, P.R. and J.W. Lorsch (1969). *Developing Organisations: Diagnosis and Action*. Reading, MA, Addison-Wesley.

Le Grand, J. (2003). *Motivation, Agency, and Public Policy: Of Knights & Knaves, Pawns & Queens*. Oxford, Oxford University Press.

Leach, S. (1982). 'In Defence of the Rational Model'. *Approaches in Public Policy*. Ed. S. Leach and J. Stewart. London, Allen & Unwin.

Leakey, R. and R. Lewin (1992). *Origins Reconsidered: In Search of What Makes Us Human*. London, Little, Brown and Company.

LeDoux, J. (1998). *The Emotional Brain*. London, Weidenfield and Nicholson.

Lewontin, R.C. (1991). *The Doctrine of DNA: Biology as Ideology*. London, Penguin Books Ltd.

Light, P.C. (1997). *The Tides of Reform: Making Government Work 1945-1995*. New Haven, Yale University Press.

Lindblom, C. (1959). 'The Science of "Muddling Through".' *Public Administration Review* **19**(2).

Lindblom, C. (1980). *The Policy-Making Process (2/e)*. New York, Prentice Hall.

Lorenz, K. (1963). *On Aggression*. London, Methuen and Co Ltd.

Mackenzie, W. and J. Grove (1957). *Central Administration in Britain*. London, Longmans.

Mansbridge, J., Ed. (1990). *Beyond Self-Interest*. Chicago, Chicago University Press.

March, J.G. and J.P. Olsen (1989). *Rediscovering Institutions and The Organizational basis of Politics*. Oxford, Maxwell Macmillan International.

Margolis, H. (1982). *Selfishness, Altruism and Rationality: A Theory of Social Choice*. Chicago, University of Chicago Press.

Marsh, D.S. and G. Stoker (Eds) (1995). *Theory and Methods in Political Science*. London, Macmillan.

McGregor, D. (1985). *The Human Side of Enterprise [1960]*. London, Penguin.

McKenzie, J. (1996). *Paradox: The Next Strategic Dimension*. New York, McGraw-Hill.

Miles, R. and C. Snow (1978). *Organizational strategy, Structure and Processes*. New York, McGraw-Hill.

Milgram, S. (1997). *Obedience to Authority [1974]*. London, Pinter & Martin Psychology.

Miller, D. (1990). *The Icararus Paradox: How Exceptional Companies Bring About Their Own Downfall*. New York, Harper Business.

Mintzberg, H. (1975). 'The Manager's Job: Folklore and Fact.' *Harvard Business Review* **75**(4).

Mintzberg, H. (1980). *The Nature of Managerial Work*. New York, Prentice-Hall.

Mintzberg, H. (1990). 'The Design School: Reconsidering the Basic Premises of Strategic Management.' *Strategic Managament Journal* **11**.

Mintzberg, H. (1991). 'Learning 1, Planning 0: Reply to Igor Ansoff.' *Strategic Managament Journal* **12**.

Mintzberg, H. (1994). *The Rise and Fall of Strategic Planning*. London, Prentice Hall.

Mintzberg, H., B. Ahlstrand, et al. (1998). *Strategy Safari: A Guided Tour Through the Wilds of Strategic Management*. London, Prentice-Hall.

Morgan, G. (1986). *Images of Organisation*. London, Sage.

Morgan, G. (1996). *Images of Organisation (2/e)*. London, Sage.

Morris, D. (1967). *The Naked Ape*. New York, Dell Publishing Co, Inc.

Norman, R. (2003). *Obedient Servants? Management Freedoms and Accountabilities in the New Zealand Public Sector*. Wellington, Victoria University Press.

Ormerod, P. (1999). *Butterfly Economics*. London, Faber and Faber Limited.

Peters, B.G. (1998). 'What Works? The Antiphons of Administrative Reform.' *Taking Stock: Assessing Public Sector Reforms*. B.G. Peters and D.J. Savoie. Ottawa, Canadian Centre for Management Development, McGill-Queen's University Press.

Peters, B.G. and D.J. Savoie, Eds. (1998). *Taking Stock: Assessing Public Sector Reforms*. London, McGill — Queen's University Press.

Peters, G.B. (1999). *Institutional Theory In Political Science*. London and New York, Pinter.

Peters, T. and R. Waterman (1982). *In Search of Excellence: Lessons From Americas Best Run Companies*. New York, Harper and Row Publishing.

Pinker, S. (1994). *The Language Instinct*. London, Penguin.

Pinker, S. (2002). *The Blank Slate: The Modern Denial of Human Nature*. London, Allen Lane.

Pollitt, C. (1984). *Manipulating the Machine*. Hemel Hempstead, Allen and Unwin.

Pollitt, C. (1990). *Managerialism and the Public Services (1/e)*. Oxford, Blackwell.

Pollitt, C., J. Birchall, et al. (1998). *Decentralising Public Service Management*. London, Macmillan.

Pollitt, C. and C. Talbot, Eds. (2004). *Unbundled Government: A Critical Analysis of the Global Trend to Agencies, Quangos and Contractualisation*. London, Routledge (forthcoming).

Pollitt, C., C. Talbot, et al. (2004). *Agencies: How Government's Do Things Through Semi-Autonomous Organisations*. London, Palgrave (forthcoming).

Pondy, L.R., R.J. Boland, et al. (1988). *Managing Ambiguity and Change*. Chichester, John Wiley.

Poundstone, W. (1988). *Labyrinths of Reason: Paradox, puzzles and the Fraility of Knowledge*. London, Penguin Books Ltd.

Price Waterhouse, C.I.T. (1996). *The Paradox Principles*. Chicago, Irwin.

Putnam, R. (1993). *Making Democracy Work: Civic Traditions in Modern Italy*. Princeton, Princeton University Press.

Quinn, J.B. (1980). *Strategies for Change: Logical Incrementalism*, Richard D. Irwin, Inc.

Quinn, R.E. (1988). *Beyond Rational Management*. San Francisco, Jossey-Bass.

Quinn, R.E. and K.S. Cameron, Eds. (1988). *Paradox and Transformation: Towards a Theory of Change in Organization and Management*. Cambridge, Ballinger Pub.

Quinn, R.E., S.R. Faerman, et al. (1996). *Becoming a Master Manager: A Competency Framework (2/e)*. New York, Wiley.

Rawls, J. (1971). *A Theory Of Justice*. Oxford, Oxford University Press.

Reynolds, V. (1976). *The Biology of Human Action*. San Fransico, W H Freeman and Company.

Rose, H. and S. Rose, Eds. (2000). *Alas, Poor Darwin: Arguments Against Evolutionary Psychology*. London, Jonathon Cape.

Rose, S., R.C. Lewontin, et al. (1984). *Not in Our Genes: Biology, Ideology and Human Nature*. London, Penguin.

Rosenhead, J.E. (1989). *Rational Analysis for a Problematic World*. New York, Wiley & Son.

Schlesinger, A. (1986). *The Cycles of American History*. Boston, MA, Houghton Mifflin.

Schneider, F. and D. Enste (2003). *The Shadow Economy*. Cambridge, Cambridge University Press.

Self, P. (1993). *Government by the Market? The Politics of Public Choice.* London, Macmillan.

Shaw, P. (2002). *Changing Conversations in Organizations: A Complexity Approach to Change.* London and New York, Routledge.

Simon, H.A. (1946). 'The Proverbs of Administration.' *Public Administration Review* 6(1): 53-67.

Simon, H.A. (1957). *Administrative Behavior (2/e).* New York, Macmillan.

Simon, H.A. (1960). *Decision Making and Organizational Design,* [reprinted in Pugh 1985].

Simon, H.A. (1999). *The Sciences of the Artificial.* Cambridge, MA, The MIT Press.

Simon, H.A. (2000). *Public Administration in Today's World of Organizations and Markets.* Washington DC, American Political Studies Association Conference.

Singer, P. (1983). *The Expanding Circle.* Oxford, Oxford Paperbacks (Oxford University Press).

Singer, P. (1999). *A Darwinian Left: Politics, Evolution and Cooperation.* London, Weindenfeld & Nicolson.

Smith, K.K. and D.N. Berg (1988). *Paradoxes of Group Life.* San Francisco, Jossey Bass.

Sober, E. and D.S. Wilson (1998). *Unto Others: The Evolutions and Psychology of Unselfish Behavior.* Cambridge, MA, Harvard University Press.

Somit, A. and S. Peterson, Eds. (2003). *Human Nature and Public Policy: An Evolutionary Approach.* New York, Palgrave.

Spann, R.N. (1981). 'Fashions and Fantasies in Public Administration.' *Australian Journal of Public Administration* **XL**: 12–25.

Stacey, R.D., G. Douglas, et al. (2000). *Complexity and Management: Fad or Radical Challange to Systems Thinking?* London and New York, Routledge.

Streatfield, P.J. (2001). *The Paradox of Control in Organisations.* London and New York, Routledge.

Talbot, C. (1995). 'Developing Strategic Managers for UK Public Services: A Competing Values and Competences Approach. *Interdisciplinary Institute of Management.* London, London School of Economics, University of London.

Talbot, C. (1996). 'Ministers and Agencies: Responsibility and Performance.' *Second Report: Ministerial Accountability and Responsibility HC313-ii.* Public Service Select Committee, II (Memoranda of Evidence) London, HMSO.

Talbot, C. (1997). 'UK Civil Service Personnel Reforms: Devolution, Decentralisation and Delusion.' *Public Policy and Administration* **12**(4).

Talbot, C. (2002a). 'A Treasury Equation.' *White Paper* **1**(1).

Talbot, C. (2002b). 'Pennies, Performance and Politics.' *New Economy* **9**(1).

Talbot, C. (2004). 'Executive Agencies: Have They Improved Management in Government?' *Public Money & Management* **24**(1).

Thain, C. and M. Wright (1996). *The Treasury and Whitehall: The Planning and Control of Public Expenditure, 1976–1993.* Oxford, Clarendon Press.

Tiger, L. and R. Fox (1971). *The Imperial Animal.* New York, Dell Publishing Co.

Treasury Committee (2002). *Spending Review 2002 Minutes of Evidence 17 July and 18 July 2002.* Andrew Dilnot, David Walton, Prof. Colin Talbot, Nicholas Macpherson (HMT), Adam Sharples (HMT), Gordon Brown (Chanecellor), Ed Balls (HMT). London, House of Commons.

Trivers, R.L. (1972). 'Evolution of Reciporocal Altruism.' *Quarterly Review of Biology* **46**: 33–57.

Tucker, J. (1967). *A Treatise Concerning Civil Government [1781]*. New York, Augustus M. Kelley, Publishers.

Whittington, R. (1993). *What is Strategy and Does it Matter?* New York, Routledge.

Wilson, E.O. (1975). *Sociobiology: The Abridged Edition*. Cambridge, MA, The Belknap Press of Harvard University Press.

Wilson, E.O. (1998). *Consilience: The Unity of Knowledge*. Little, Brown and Co.

Wilson, J.Q. (1997). *The Moral Sense*. London, Simon and Schuster.

Wilson, P. (1988). *The Domestication of the Human Species*. New Haven and London, Yale University Press.

Winston, R. (2002). *Human Instinct*. London, Bantam Books.

Wright, L. (1998). *Twins: Genes, Environment and the Mystery of Human Identity*. London, Phoenix.

Wright, R. (1995). *The Moral Animal: Why We Are the Way We Are. The New Science of Evolutionary Psychology*. New York, Vintage Books.

SOCIETAS: essays in political and cultural criticism

Vol.1 Gordon Graham, *Universities: The Recovery of an Idea*
Vol.2 Anthony Freeman, *God in Us: A Case for Christian Humanism*
Vol.3 Gordon Graham, *The Case Against the Democratic State*
Vol.4 Graham Allen MP, *The Last Prime Minister*
Vol.5 Tibor R. Machan, *The Liberty Option*
Vol.6 Ivo Mosley, *Democracy, Fascism and the New World Order*
Vol.7 Charles Banner/Alexander Deane, *Off with their Wigs!*
Vol.8 Bruce Charlton/Peter Andras, *The Modernization Imperative*
Vol.9 William Irwin Thompson, *Self and Society* (March 2004)
Vol.10 Keith Sutherland, *The Party's Over* (May 2004)
Vol.11 Rob Weatherill, *Our Last Great Illusion* (July 2004)
Vol.12 Mark Garnett, *The Snake that Swallowed its Tail* (Sept. 2004)
Vol.13 Raymond Tallis, *Why the Mind is Not a Computer* (Nov. 2004)
Vol.14 Colin Talbot, *The Paradoxical Primate* (Jan. 2005)
Vol.15 J.H. Grainger, *Tony Blair and the Ideal Type* (March 2005)
Vol.16 Alexander Deane, *The Great Abdication* (May 2005)
Vol.17 Neil MacCormick, *Who's Afraid of a European Constitution* (July)
Vol.18 Larry Arnhart, *Darwinian Conservatism* (September 2005)
Vol.19 Henry Haslam, *The Moral Mind* (November 2005)
Vol.20 Alan and Marten Shipman, *Knowledge Monopolies* (January 2006)
Vol.21 Kieron O'Hara, *The Referendum Roundabout* (March 2006)
Vol.22 Paul Robinson, *Doing Less With Less: Britain more secure* (May '06)

Public debate has been impoverished by two competing trends. On the one hand the trivialization of the media means that in-depth commentary has given way to the soundbite. On the other hand the explosion of knowledge has increased specialization, and academic discourse is no longer comprehensible. As a result writing on politics and culture is either superficial or baffling.

This was not always so — especially for politics. The high point of the English political pamphlet was the seventeenth century, when a number of small printer-publishers responded to the political ferment of the age with an outpouring of widely-accessible pamphlets and tracts. Indeed Imprint Academic publishes facsimile C17th. reprints under the banner 'The Rota'.

In recent years the tradition of the political pamphlet has declined—with most publishers rejecting anything under 100,000 words. The result is that many a good idea ends up drowning in a sea of verbosity. However the digital press makes it possible to re-create a more exciting age of publishing. *Societas* authors are all experts in their own field, but the essays are for a general audience. Each book can be read in an evening. The books are available retail at the price of £8.95/$17.90 each, or on bi-monthly subscription for only £5/$10. Details: **imprint-academic.com/societas**

IMPRINT ACADEMIC, PO Box 200, Exeter, EX5 5YX, UK
Tel: (0)1392 841600   Fax: (0)1392 841478   sandra@imprint.co.uk
imprint-academic.com/societas